NURSERY
EARTH

NURSERY EARTH

The Hidden World of Baby Animals
and the Amazing Ingenuity of Life

DANNA STAAF

Foreword by Richard Strathmann, PhD

THE EXPERIMENT
NEW YORK

The Experiment, LLC
220 East 23rd Street, Suite 600
New York, NY 10010-4658
theexperimentpublishing.com

THE EXPERIMENT and its colophon are registered trademarks of The Experiment,
LLC. Many of the designations used by manufacturers and sellers to distinguish their
products are claimed as trademarks. Where those designations appear in this book and
The Experiment was aware of a trademark claim, the designations have been capitalized.

The Experiment's books are available at special discounts when purchased in bulk for
premiums and sales promotions as well as for fund-raising or educational use. For
details, contact us at info@theexperimentpublishing.com.

The Library of Congress has cataloged the earlier edition as follows:

Names: Staaf, Danna, author.
Title: Nursery earth : the wondrous lives of baby animals and the
  extraordinary ways they shape our world / Danna Staaf.
Description: New York : The Experiment, [2023] | Includes bibliographical
  references and index.
Identifiers: LCCN 2023000439 (print) | LCCN 2023000440 (ebook) | ISBN
  9781615199327 (hardcover) | ISBN 9781615199334 (ebook)
Subjects: LCSH: Animals--Infancy. | Animal behavior. | Parental behavior in
  animals.
Classification: LCC QL763 .S735 2023 (print) | LCC QL763 (ebook) | DDC
  591.3/9--dc23/eng/20230127
LC record available at https://lccn.loc.gov/2023000439
LC ebook record available at https://lccn.loc.gov/2023000440

ISBN 978-1-891011-71-9
Ebook ISBN 978-1-61519-933-4

Cover design by Beth Bugler
Text design by Jack Dunnington
Cover photograph courtesy of Chester Zoo

Manufactured in the United States of America

First paperback printing May 2024
10 9 8 7 6 5 4 3 2 1

CONTENTS

FOREWORD

by Richard Strathmann

In folk tales, the young set out to make their fortunes; our interest is held by the challenges they meet and overcome. So it is with the adventures of Earth's animal babies—but with stories more varied, strange, and surprising than in the tales of those youthful heroes.

Some babies, however different from adults, are familiar: the tadpole of the frog, the caterpillar of the butterfly. But there are many more that are so unlike their adult forms that they were first described by biologists as newly discovered animals, as with the pluteus of the sea urchin and the nauplius of the barnacle. Danna Staaf reveals this wondrous world of baby animals. Few among us—amateur nature lovers and specialists alike—know their extraordinary diversity. Staaf has brought together biologists' studies from land, ocean, lakes, and streams, for the first time in popular book form. The varied devices by which young life meets the world's threats and obstacles astonish and entertain. *Nursery Earth* shows us baby animals as creatures unto themselves—not just steps on the way to becoming adults but amazing living things in their own right.

All creatures that are composed of many cells, including us humans, benefit from reverting to a single cell as part of our life cycle. Passing through this one-cell bottleneck provides benefits—it purges the future generation of harmful mutations and pathogens and makes sexual reproduction possible—but it also creates a challenge: how to start as an egg and survive and grow to full multicellular capabilities in a world of risks. The next generation depends on animals' meeting this challenge of a precarious one-cell beginning. From a poorly equipped start, each developing animal requires nutrition, defense, oxygen, and waste disposal, plus, often, a way to move about. *Nursery Earth* tours us through the fascinating and varied ways that these needs and others are met.

Staaf describes the wonder of seeing a little sphere—an egg—turn into a functioning animal in the lab—and the even greater wonder of how many such babies survive in the harsher world outside the lab. Some hazards are old: predators, parasites, pathogens, desiccation, suffocation. Some hazards are peculiar to particular kinds of reproduction: cannibalism by siblings, transport by currents beyond habitable environments, even being eaten from the inside after being injected with another animal's eggs. Staaf also speaks of growing concerns for the future. Some manufactured chemicals are new hazards. Old hazards change, too, as humans change Earth's environments by carbon emissions and harvesting. The perils can be formidable, but in Staaf's account, humor outweighs the pathos—parasitic wasps, a possible exception.

In *Nursery Earth*, you'll also discover the ways we're part of the story. What we humans do to care for our babies resembles strategies that are not limited to our closer kin: Mother mammals, for example, aren't the only mothers who nourish babies internally through their blood or externally through their milk. But Staaf also shows that we don't need to depend on finding humanlike similarities in other animals in order to love and admire them; instead, animals with strangely different forms and habits earn our appreciation and sympathy. There's charm in a scarab beetle sequestering its baby in a protective and nourishing ball of dung. We can admire the versatility with which some wasp embryos and starfish larvae multiply themselves.

We all, as readers, will find much to ponder and enjoy in animals' wide range of journeys from egg to adult. We and our world depend on the ways they stay alive.

RICHARD STRATHMANN, PhD, is an expert in the diverse patterns of animal development, with a particular focus on marine animals. He finds the beauty and variety of changes from eggs through embryos, larvae, and metamorphosis endlessly entertaining. He joined the faculty of the University of Washington in 1973.

INTRODUCTION

A World of the Babies, by the Babies, for the Babies

There was a child went forth every day
And the first object he look'd upon, that object he became,
And that object became part of him for the day or a certain
    part of the day,
Or for many years or stretching cycles of years.

—Walt Whitman, *Leaves of Grass*

Baby animals are undeniably cute: puppies tumbling over each other, joeys peeking out of kangaroo pouches, ducklings paddling in a wobbly line. Baby animals are also incredibly bizarre: Moth larvae mimic both snakes and feces, featherless finch chicks beg with beaks like Mondrian paintings—and let's not forget baby humans, with our squishy skulls and taste buds on our tonsils.

Baby animals are very sensitive to the environment: Bird embryos die when their eggshells are thinned by DDT or smothered by spilled oil. Shellfish larvae can't grow into edible adults until they find the perfect home in a sea full of increasingly imperfect habitat. But baby animals are also powerful enough to change the environment, for both good and ill from the human perspective. Agricultural pests like rootworms and borers are actually infant insects, and they devastate crops on every continent. At the same time, numerous beetle larvae can actually digest plastic, offering hope for pollution cleanup.

Even human babies embody this combination of vulnerability and voracity. When we're born, we don't know a single language, leaving us at the mercy of our adult caretakers. Yet we are so ravenous to learn that we can pick up as many languages as we are exposed to, an ability the adults around us have lost. Similarly, as babies, our incomplete immune systems put us at risk from infections that rarely affect adults. But these same baby immune systems are primed to build relationships with beneficial bacteria that will serve us the rest of our lives. If we're exposed to too many dangers or deprived of necessary resources, our sensitivity becomes a weakness, but in the right environment, it blooms into a superpower.

Scientists who study early life stages, by dripping chemicals on tadpoles or feeding chicks experimental diets or injecting genes into fly eggs, are known as developmental biologists. Developmental biology is the study of how animals build their bodies—an examination of every process between fertilization and maturity. As a field of research, it has progressed through its own fascinating and sometimes turbulent developmental stages. In the nineteenth century, it was called embryology, and its practitioners peered through microscopes to watch an egg cell cleave into two, four, eight, and many more cells, then eventually grow a gut and a brain. The work of embryologists expanded to produce and bud off the entire field of genetics in the twentieth century, and developmental biology is now expanding again to link the microscopic stages of animal growth with global environment and ecology.

Unexpected connections are the forte of developmental biology. "Adult-onset" diseases like cancer and diabetes are increasingly understood to result from influences in early life, even as early as the womb. Not only humans but many other animals face challenges to health, longevity, and survival that trace their roots to chemical exposure or resource limitation in babyhood. And although these challenges highlight the vulnerability of youth, they also illuminate its adaptability. When a developing baby doesn't get enough nutrition, it can preferentially devote its limited energy to growing critical organs like the heart and brain, leaving more redundant organs like kidneys to suffer the brunt of starvation. This postpones problematic symptoms until later in life, giving the animal a chance to reproduce first. Thus, surviving long enough to grow up and manifest kidney disease is a triumph of the flexibility of early development.

At no other stage in our lives are animals more capable of perceiving and responding to changes in the environment. In intimate conversation with our inanimate surroundings as well as with our fellow cohabitors of Earth, we mold our bodies to match our world. Within this capacity for change lies the future of life as we know it.

The diversity of developmental forms

When we think of the world's diverse animal life, we usually think of adult animals: frogs and butterflies, jellyfish and echidnas. Not tadpoles and caterpillars, ephyra and puggles. (Believe it or not, those baby names match the parents that precede them.) Even when we turn our minds to baby animals, we tend to forget that they don't always look like their parents. This leads to the occasional amusing contradiction in children's books, like a "daddy caterpillar" or a "baby bee." In reality, daddy caterpillars are moths or butterflies, and baby bees are wingless white larvae.

What *is* a larva? (Besides the singular form of larvae.) The word describes a baby that goes through a distinct metamorphosis to become an adult. Most animals have one or more larval stages. Because larvae can look so different from adults, this creates a constant puzzle for biologists, who must piece disparate forms into a single life cycle. Some species' larvae have never been seen. Other larvae are not yet associated with an adult.

Larvae don't have to be tiny, nor are all tiny babies necessarily larval forms. Bluefin tuna and kangaroos, both of which can grow to well over 3 feet (1 m) as adults, produce babies about an inch (2.5 cm) long. The tuna hatchling is a larva (see insert, photo 1), the kangaroo joey is not. On the other end of the scale are the chick of a kiwi (not a larva) and the maggot of a tsetse fly (a larva), both born nearly the same size as their parents.

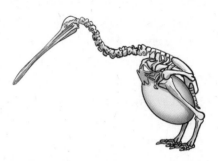

This illustrated X-ray view into a mother kiwi makes it obvious that she lays the largest eggs of any bird, relative to her body-size. A bellyful of yolk sustains the chick until it learns to forage for food.

This size variation arises because animals can't invest infinite energy in producing offspring. There's a trade-off between size and number. Kiwis lay one gigantic egg at a time, while tuna spawn millions of mini eggs. In both cases, the babies that hatch from the eggs are independent, striking out on their own. Kiwi parents allocate their reproductive effort to building mass, producing a baby large enough to have a good chance of solo survival. Tuna parents allocate theirs to quantity, producing enough babies that it doesn't matter if only a few survive.

Kangaroos follow a third route, pouring their resources into parental care. Although each baby is minuscule, it receives the warm protection of a pouch and a constant infusion of milk for up to a year after birth. Thus, although a newborn kangaroo is as tiny as a newborn tuna, an *independent* kangaroo is far closer in size to an adult.

Whether a baby turns to its parents or to the wild world to supply its needs, it is exquisitely well-equipped to do so. Kangaroo joeys have tough arms for climbing from the birth canal to the pouch. Larval tuna have (relatively) enormous jaws for swallowing prey nearly as big as they are. Larval parasites may be some of the most specialized babies of all, built to create links between

Newborn kangaroos look like early embryos but crawl better than newborn humans. They make their own way into the pouch, where they nurse and grow until they're more than half their parents' height.

disparate forms of life. As the tiny seeds of mighty trees are adapted to find a new home by hooking onto an animal carrier or hitching a ride on the wind, so baby tapeworms help themselves travel to a new host. A tiny "worm seed" can infect a human through the consumption of undercooked pork, as cleverly described by the embryologist-poet Walter Garstang:

> He's very small, a mere pin's head, beset with six small hooklets,
> Is whirled about by wind and rain through puddles, fields and
>   brooklets;
> But if a pig should swallow him, as many porkers do,
> He's made a start with no mistake: he's on the road to you![1]

A tapeworm baby needs to be eaten in order to complete its life cycle, a typical parasite feature. As it passes from host to host, a single parasite can infect a wide range of animals in a wide range of habitats, connecting snails to birds and wetlands to forests. Similar connections are made even by animal babies that die when they're eaten. Young life-forms are easier prey than their parents, accessible to a greater range of predators. Many hungry animals depend for their meals on the profusion of progeny produced by their fellows.

The hidden abundance of youth

For many species, babies comprise the majority of their life cycle. Most animals on Earth are, in fact, babies.[2]

It might be easiest to understand this as an ephemeral spring-time truth, since we're used to seeing one duck parent trailed by a dozen fuzzy offspring. The children's classic *Make Way for Duck-lings* contains four times as many babies as adults. In March, a square meter of water (about 11 sq ft) in a North Carolina pond can hold fifteen thousand tadpoles, the product of breeding by only a few hundred adults.[3] In both cases, these babies grow to adult-hood in a matter of weeks, and for the rest of the year, no ducklings or tadpoles can be found. So we might conclude that only during certain limited times are adults outnumbered by babies. However, as climate change brings spring weather earlier and earlier, the breeding season of many species is extended, and so is the period of time during which babies rule the roost.[4, 5]

What's more, other animals often linger much longer in child-hood and youth. Salmon are born in fresh water and live in streams for up to two years as fry before they mature and move out to the ocean. Many types of salmon spend no more time as seafar-ing adults than they did as river-dwelling babies. Given the natural population attrition over time, as salmon are eaten by predators or succumb to parasites, the total number of fry is typically greater than that of adults at any time of year.[6]

Although salmon babies stick to rivers, the ocean serves as a giant nursery for uncounted other species. Surface waters froth with billions of baby fish, squid, crabs, and more, and it seems like each new expedition to the deep sea uncovers another astonishing cradle of life. In 2021, in Antarctica's Weddell Sea, cameras towed at hundreds of meters' depth revealed an icefish breeding colony of 93 square miles (240 sq km) filled with an estimated 60 million nests. The average number of eggs per nest was 1,735, making the total number of babies in this nursery well over a hundred billion.[7]

This previously unknown jackpot of baby fish revised our whole understanding of the Antarctic ecosystem.

As for humans? Currently, about 22 percent of the world's *Homo sapiens* population is under the age of eighteen—likely the lowest this percentage has ever been. Less than a century ago, it was 31 percent.[8] The relative proportion of children and teens varies geographically, with only 17 percent of the Japanese population under the age of twenty, while 60 percent of the Nigerien population falls into this category.[9] On average, adult humans outnumber children, but in many places—from the country of Niger to any schoolyard—the reverse is true.

Including humans with other animals can be a touchy subject. Merriam-Webster offers multiple definitions for "animal": first, *any of a kingdom (Animalia)*, and second, *one of the lower animals as distinguished from human beings*. Both definitions have their uses. The species *Homo sapiens* belongs indisputably to the order Primates, phylum Chordata, kingdom Animalia, a fact encompassed in the first definition. At the same time, *Homo sapiens* is the only species that engineers global-environment-altering technology, and engages in moral debates about said technology (among many other topics). The second definition allows us to refer to all the animals that *don't* do this with a single word.

However, our developmental biology illuminates our kinship with the rest of the kingdom. As a human embryo, I looked pretty fishy for a while. I also at times resembled a reptile and a chick. The similarities are eye-catching enough that some early biologists encoded them in law, contending that each animal displays the evolutionary history of its species over the course of its development. We now know, as we'll explore further in chapter 8, that this superficially compelling "law" fails to capture the true intersection of development and evolution, but shared embryonic features still inform our understanding of relationships between animals—including humans. While this book is not about human development (many other excellent texts are available on that subject), our species will come up from time to time.

After all, I am a human, and you most likely are one, too. I, like you, began life as a baby. Also probably like you, I don't remember it. I know that I depended on my parents and other caregivers, and I remain grateful to them for keeping me alive. I know that I was complicit in the process, crying for the attention I needed, producing attractive facial expressions and postures to garner care. I absorbed environmental input, both actively when I put dirt in my mouth and passively as I experienced the hot summers, poor air quality, and minimal rainfall of Los Angeles in the late twentieth century. I was fortunate to be given consistent affection and nutrition and to be brought periodically to a beach where I could enrich my sampling of dirt and dry grass with salt water and sand.

How much of who I am today is shaped by the genes in my mother's egg cell and my father's sperm cell, and how much by my experiences from the womb onward? This age-old question of nature versus nurture sits at the very heart of developmental biology.

From field to lab and back again—the development of developmental biology

We are as much a product of our environments as of our genes. Your immune system, your digestive system, even your brain and your bones all develop with environmental input, whether that input is bacteria, exercise, or diet.[10] Within the rest of the animal kingdom, bacteria make it possible for insect embryos to grow, amphibian eggs to hatch, and squid to mature. Temperature determines the sex of certain turtles; location, the sex of certain worms. Diet dictates a bee's caste; predators make water fleas grow spines. Birds and mammals grow bones and muscles in response to gravity and stress. Larvae of all kinds metamorphose in response to temperature, light, texture, or smell.

Many of these intimate connections between development and environment were already understood more than a hundred years ago. But in the middle of the twentieth century, when the discovery of DNA allowed scientists to begin reading and manipulating genetic code, the study of development moved out of the natural environment into a carefully controlled laboratory setting. Developmental biologists began to not only disregard but actively avoid environmental influences, so they could focus on the results of genetic modifications.

Biology underwent a schism. On one side were ecologists, studying organisms and networks of organisms in their natural habitat. On the other were geneticists, focused on the molecules that constitute and create life. Neither was wrong about the importance of their work, but they had such difficulty talking to each other that many university biology departments divorced, engendering two separate departments and lots of ruffled feathers.

This is what I discovered as a seventeen-year-old college freshman, newly fledged from my nest in Los Angeles and settling in to study marine biology at the University of California, Santa Barbara. My interests and classes aligned mostly with the Department of Ecology, Evolution, and Marine Biology. But if I wanted to learn about embryos, I had to take a class in the opposing Department of Molecular, Cellular, and Developmental Biology. The separation of marine and developmental biology is especially ironic, as the embryos of marine animals like sea urchins catalyzed numerous foundational breakthroughs in developmental biology.

However, none of these marine species became true model organisms—they were too tricky to raise in the lab. Instead, an enormous body of research was built around the laboratory study of genetic inheritance and mutations in six nonmarine model organisms: the fruit fly, the roundworm, the mouse, the chicken, the frog, and the zebra fish. Scientists picked these species because they're easy to raise and require minimal environmental input, and studies on them have yielded huge discoveries, from the "toolkit genes" that organize development across the entire animal kingdom to developmental genes that are implicated in human cancer.

However, most animals, including humans, are not model organisms. Most of us could not develop successfully to adulthood in a sterile laboratory environment. Even the model organisms themselves can have trouble—scientists raised "germ-free" mice in the complete absence of bacteria, and discovered a suite of metabolic, neurological, and behavioral abnormalities. As for factors like temperature and diet, those would never be as constant in the real world as they are in the lab. The same precisely optimized nutrition is used to raise chicks in universities around the globe.[11] What would happen if these animals were reared in more realistic conditions?

"Will the egg be computable?" asked Lewis Wolpert, a widely respected developmental biologist, in 1994. Which was to say, given complete knowledge of every molecule in a fertilized egg, could its development to adulthood be predicted? By 2006, Wolpert asserted that it would happen "in the next fifty years." However, nearly everything that we have learned about developmental biology since then points in the opposite direction.

We used to think of development as a computer program. It has revealed itself instead as a collaborative, continuous performance.[12] Development is not merely how we build ourselves, it's how the world builds us. Rather than a rigid set of instructions defined by genes, it's a set of possibilities influenced by and adapted to the environment.

Slowly, environmental concerns have forced their way back into our understanding of development. In 1982, a group of scientists realized that attempts to help endangered sea turtles by incubating eggs at controlled temperatures were actually harming the population by producing babies of all one sex.[13] At first, such cases seemed like exceptions. But as the evidence mounts, we begin to understand that the development of *all* organisms is defined by their environment.

Studying a greater variety of organisms in a greater variety of conditions becomes increasingly important. In 2003, researchers discovered that tadpoles are more vulnerable to damage from pesticides in an environment that also contains predators—which is, let's be honest, nearly every environment that tadpoles face. As the lead author wrote, "it is the lethality of pesticides under natural conditions that is of utmost interest."[14] Similarly, tadpoles exposed to different pesticides that may not seem to harm them directly suffer a reduced immune response, leaving them more vulnerable to parasites than they would otherwise be.[15] Laboratory? No parasites. Real world? Crawling with them.

Not only dangers but also opportunities that would never arise in the lab abound in the real world. Toad eggs in northern India have been found developing in small "ponds" that are actually rain-filled elephant footprints.[16] Before witnessing such a specific connection, we might have expected no relationship between toads and elephants—neither preys on the other, and they don't compete for resources. But when we take each organism's entire life cycle into account, the links between species rapidly multiply. Do declines in elephant numbers hurt toad populations, by limiting their nursery sites? No one has yet gathered data on this, or myriad other potential developmental dependencies.

The connections between development and environment illuminate the enduring unity of biology, separation of departments notwithstanding. Microscopic molecules record and store information about the environment as animals encounter it, and the animals use this information to build bodies and behavior best suited to their world. Cells aggregate to produce the bodies of larvae and adults, which aggregate to fill ecosystems. Both individual organisms and entire food webs change continuously with the cycling seasons. We think of seasons as wet or dry, hot or cold, but animals also create their own breeding seasons, spawning seasons, growing and dying seasons. The molecules inside caterpillars determine the timing of butterfly emergence season, which becomes a season of abundant food for birds, which is recorded in the molecules inside their developing chicks. Development shows us that the links are nowhere broken, as we shift in scale from molecule to ecosystem. Research on animal babies is facilitating the collaboration across scales that has become crucial to understanding and caring for our world.

An elusive profusion of squid eggs

Like most children, I adored baby animals from an early age. I bonded deeply with a pet kitten; I campaigned (unsuccessfully but perennially) for a puppy. In the margins of my school notebooks, I doodled fluffy little bodies with huge heads and eyes. Even in college, when my biology classes began introducing me to a profusion of larval forms, if I'd picked up a book about "animal babies," I would have expected it to focus on cute critters like ducklings and baby bunnies.

In this book, we *will* encounter ducklings (some are parasitic!) and bunnies (eating their mom's poop!), but we will also find many far stranger babies. Caterpillars, grubs, larvae of all kinds—these babies may be less adorable, but they are no less important. What they lack in immediate visual appeal they make up for in fantastical anatomy, behavior, and transformations. These forms link the ecosystems of today and build the biology of tomorrow.

It was after college that a fascination with the true diversity of animal babies fully gripped me. Beginning my pursuit of a graduate degree in a laboratory focused on the biology of Humboldt squid, *Dosidicus gigas*, I learned that no one had ever seen the eggs of this species in the wild. The same could be said for plenty of animals, but the Humboldt squid is not the kind of small, rare, or endangered species for which you might expect this to be the case. No,

Humboldt squid grow up to 6 feet (2 m) long, swim through both northern and southern hemispheres of the eastern Pacific Ocean, and support the largest squid fishery in the world. People catch and eat nearly a billion tons of Humboldt squid every year.[17] Humboldt squid, in turn, catch and eat countless fish, crabs, and fellow squid. An animal like this has got to be making plenty of babies. So where were they?

I set out to study the early life stages of Humboldt squid, and because I actually wanted to finish my degree someday, I had to do more than hope that I could find babies in the open ocean where no one had before. When scientists want to study babies that aren't readily collected in the wild, we try to make them ourselves. In 2006, on a research ship in the Gulf of California, a generous colleague taught me how to collect eggs and sperm from adult squid and conduct *in vitro* fertilization.

The process began after dark, when Humboldt squid rise from the depths to the surface and are more easily caught. Often, it was past midnight by the time I had both eggs and sperm isolated in glass dishes. Then I would stay up for hours more, carefully mixing, changing water, watching for signs of successful fertilization and attempting to match the eggs' early development to reference drawings of other squid species (no one had ever published on Humboldt squid development before). The results were encouraging, but hardly thrilling—many eggs failed to develop properly, while others fell prey to fungal or bacterial infection.

During the day, I napped.

One afternoon, near the end of our two-week research cruise, my bunkmate burst into the room and woke me excitedly. "Guess what, Danna!"

I guessed, blearily and correctly: "You found an egg mass!"

Several researchers had been diving regularly throughout the trip. Since we were operating over deep water, far from shore, they had to be specially trained for "blue-water diving"—safely exploring the open ocean in the absence of rocks, kelp, sand, or any structure at all other than the boat itself. On these dives, they had found jellies of many kinds, most small enough to be scooped up in a collecting jar. On this day, however, they had encountered an enormous gelatinous mass.

They had taken video as well as samples. All of us scientists crowded around a TV in the tiny ship laboratory to stare at a blob so large and diffuse a diver could swim through it. It was transparent, studded with tiny embryos like stars. On the screen, a diver reached out to fill an open jar with embryos and their surrounding jelly. In the lab, I was ecstatic to receive one of these jars for my own. I had been struggling to produce babies artificially, and here were the first naturally produced babies ever to be found in the wild!

They began hatching that very day as perfectly formed little squid, smaller than rice grains (see insert, photo 2). Their eyes were huge relative to their body size, just like a human baby's, and so was their funnel—the tube a squid uses to breathe and swim. These proportions triggered cuteness-recognition algorithms in my human brain, and I felt compelled to care for the squid babies assiduously.

The next day, we disembarked from the ship in Sonora, Mexico, the home of several of our science crew. Most of the babies stayed with them, while I puzzled over how to bring a small sample of hatchlings on the flight back to my lab in Monterey, California.

This was 2006, when you were still allowed to carry a water bottle onto an airplane without emptying it first. I filled my bottle with seawater and eight squid babies, then walked through security in the Guaymas airport, feeling the illicit thrill of smuggling with none of the actual danger. All the hatchlings survived the trip, and I spent the next week studying their swimming behavior and coaxing them unsuccessfully to eat.

We also extracted DNA from a frozen sample and confirmed that the egg mass indeed belonged to a Humboldt squid. (There had been little doubt, but since we hadn't observed an adult laying

the eggs, we couldn't be positive without genetic identification.)
Estimating the size of the mass from the video and the density of
eggs in the mass from the jars, I calculated that the whole thing
contained between half a million and two million eggs. Such a vast
number sounds incredible, but other scientists had counted tens of
millions of eggs in Humboldt squid ovaries. We reasoned that one
mother could lay a dozen similarly sized masses in her lifetime.[18]

Such profligacy! How was it that no one had ever spotted one
before? Eventually, our *in vitro* work helped us understand why.
Humboldt squid develop from fertilized eggs to swimming hatchlings in a week or less, so the egg masses are incredibly transient.
Furthermore, as the divers found, these masses hover at a depth
that's invisible from the surface. Unless you happen to dive right
next to one, you would never see it.

Yet without these nearly invisible, transitory masses, the largest
squid fishery in the world would cease to exist, and a dominant
predator of the eastern Pacific would be gone. Throughout the
world, animal babies are hidden threads tying together all the planet's ecosystems more tightly than we realize. They are now growing up in the most interconnected, rapidly changing environment
that any generation has ever experienced. It's time to pay attention
to them.

Metamorphosis beyond metaphor

It's tempting to think of development as a process with an end
product. But what is the end product? Is it the baby, caterpillar, or
larva that emerges from an egg or a womb? In many cases, there
is a distinct transition from subsisting off yolk or placenta to independent movement and feeding. In other cases, there is not, and
the organism that emerges carries yolk with it for days, continuing
to live off that parental investment, and not moving actively any
more than it did in the egg. Where then to draw the line between
process and product?

Perhaps the adult is the end product. For metamorphosing species, this is also a clear demarcation. The newly emerged moth is
no caterpillar, and it will live as a moth for the remainder of its life.
But what about the rest of us? When does a human become an
adult? Is it the onset of menstruation, the change in voice, the first
credit card? We use different rubrics for different purposes, and
we seem to think there is some nebulous point at which adulthood
has "arrived." However, development never stops; we are a work in
progress for all of our lives. Our brains and bodies keep changing.
Is menopause a developmental process? How about balding?

One butterfly scientist I spoke with emphasized how much
we love the metaphor of metamorphosis because we all want to
be able to change yet remain ourselves. I reflected that it doesn't
have to be merely a metaphor. We can change; in fact, we can't not
change. When I gave birth to my first child, the experience felt like
a rebirth of my own self. I had been an adult human before. Now
I was an adult parent. I had changed, physically and mentally and
irrevocably, both over the months of pregnancy and at the time of
delivery. It felt like an almost uncanny parallel to pupation (what
happens inside a cocoon or chrysalis) and *eclosion* (the moment of
emergence).

Metamorphosis, a phenomenon we'll explore in more depth in
chapter 9, is a feature that allows a single organism to build itself
multiple bodies, each adapted to the demands of a specific environment. Aquatic insects like mayflies lose their childhood gills and
grow wings to carry them on mating flights. Frogs resorb the tails
of their watery youth and sprout legs to hop and climb.

It may seem precarious for a species' survival to depend on two
separate habitats. There's also a distinct advantage: Babies that live
in different places and eat different food from their parents experience no competition with the older generation. Adults can consume
as much as they need without taking resources away from their children. This is especially useful when certain ravenous children grow

to the same size as adults—or even larger. Adult Goliath beetles are some of Earth's biggest insects, and they are vegetarians, feeding on fruit and tree sap. Their larvae, however, grow up to *twice* the weight of an adult on an as-yet-unknown diet that likely includes significant protein, suggesting a predatory habit. (In captivity, they readily consume cat kibble.[19]) Paradox frogs, also known as shrinking frogs, engage in the opposite dietary shift. They grow to an enormous size as algae-eating tadpoles, only to metamorphose into much smaller adults that prey on insects.[20]

All animals move between environments as they mature, whether it's a matter of scale (from the miniature environment of a tiny fish larva to the oceanic environment of a full-grown tuna) or location (from the saltwater habitat of young eels to the freshwater homes of adults) or, very often, both. Some larvae can even be swept like dandelion seeds across continents and seas, connecting far-flung locales and sprouting new populations.

Humans fall into all of these categories, too. Some of us stay in our hometowns and scale up from a playground to an office building, while others emigrate thousands of miles. Wherever we are and however far we travel, we share every environment on the planet with an enormous diversity of animal life.

The young members of all species are active participants in our planet's drama. They are consumers and producers, competitors and cooperators. The world's most destructive crop pests are baby moths and beetles, while some of the most effective methods of pest control are baby wasps. Many endangered species are at risk, not because of threats to the adults but because their babies are running out of habitat or suffering from pollution. And speaking of pollution, larval invertebrates play a crucial role in the study of chemical toxicity, their sensitivity helping us make decisions for our own safety and that of our environment—which are, after all, inextricably linked.[21]

Babies as links across space and time

In the pages to come, we'll explore animal development from egg to metamorphosis, witnessing the intimate interdependence between each form and its environment. We'll see how important these early life stages are to the ecosystems they inhabit, and how vulnerable to perturbations like pollution and climate change. The field known as *ecological developmental biology* has expanded rapidly in recent years, with incredible findings since 2020 alone.

Astonishing research has illuminated how babies connect different parts of the world, from fish eggs tough enough to survive traveling between lakes in duck intestines[22] to sea star larvae that clone themselves as they surf ocean currents in a reverse of Vasco de Gama's famous voyage.[23] Bold and sturdy adventurers, babies also face huge risks, like the warming water that makes baby sharks more visible to predators[24] and the inbreeding that kills endangered bird chicks long before they can hatch.[25] Perhaps most important of all, we are learning more about how microbes guide animal growth. The word *microbe* is a shortening of *microbios*, or "tiny life." The "micro" part implies that microbes are living things that can only be seen with a microscope—but does that make any microscopic embryo a microbe? Nobody would argue for that, since the embryo's small size is just a phase. And yet, bread mold that can grow large enough for us to see with the naked eye *is* considered a microbe. Go figure. When I talk about microbes in this book, I'll be referring mostly to single-celled bacteria, fungi, and viruses. Insect larvae cooperate with microbes like these in their gut to break down plastics,[26] and a mother's vaginal microbes can stabilize the health of human babies born by C-section.[27]

In addition to creating links across space, animal babies also illuminate connections over time. Development offers a window into the history of life on Earth. The gills of insect larvae evolved into the wings of adults. Neoteny, or retaining childlike features, is a part of how we humans evolved from other apes. Life began with single-celled organisms, and embryos daily demonstrate the wonder of building one cell into multicellular life.

In fact, when you consider that the single-celled ancestors of animals simply replicated themselves, it's incredibly weird that we make babies at all. Plenty of multicellular animals can still pull off asexual replication: Anemones bud clone after clone to fill tide pools. Flatworms break into pieces, each of which becomes a whole worm. So why don't we all reproduce that way? As it turns out, starting the next generation with a single, fragile cell provides surprising advantages, from clearing out disease to promoting cooperation within the new body.

Most people don't talk about miracles much, but one place the word seems to crop up again and again is in reference to a new baby. "It's a miracle," people say, in awe at the arrival of a new human, or if they're kindly inclined toward the rest of our kingdom, a new calf or chick or puggle. Researchers who study development experience no less awe—perhaps even more, as they uncover layers of marvelous detail. Many of the scientists I spoke with for this book are parents; all were once babies themselves. Again and again I heard them express wonder at the developmental process. Is this what drives so many biologists to poetry, or are the poetically minded drawn to the science of development? In the chapters to come, we'll encounter verses by eminent researchers as well as words by full-time poets that speak to the topics at hand.

We say "it takes a village to raise a child" and we consider ourselves responsible, as a society, for the care and safety of all children. We build public schools and playgrounds and institute child protection systems. What if we broadened this view to include all the world's young? Human or hyena, squid or scorpion—we were all young once. This book will show you why that matters.

—

PART I

BUNDLES OF JOY

1

EGGS

Not Just a Bird Thing

To make one me, you just add
half of mom and half of dad—
that is what I once believed.
But now I know that I was wrong
you gave so much to me, Mom,
besides one half a set of genes . . .

—Adam Cole, "A Biologist's Mother's Day Song"[1]

"Hey, Mom, did you know it takes an elephant cell two years to become an elephant?" My eight-year-old is reading a children's book about reproduction. This particular fact puzzles me for a moment.

"Oh, do you mean a fertilized egg?" I ask.

"It just says a cell."

She shows me a page full of cartoon mammals announcing the length of their pregnancies with, "It takes an elephant cell about two years," and "It takes a rabbit cell about three weeks."[2] "I can understand why they used the word *cell*; it's short and simple," I tell her. "But it's a bit misleading because you can't grow an elephant from just any elephant cell. What they mean is a fertilized egg cell."

At that point, my husband speaks up from the kitchen. "What's so special about a fertilized egg cell?"

Eggs are all around us. My family has chicken eggs in a carton in the fridge, ready to scramble or bake into muffins. We have song-bird eggs under our eaves every spring. Beneath any stone or stick in the yard, we're likely to find a sack of spider eggs. Based on the number of baby lizards around, we know there must be lizard eggs, even though we've never been able to spot them. And inside my ovaries and those of my daughter are hundreds of thousands of human eggs.

An egg is one of the most incredible biological products in the world. It's a nearly complete package of everything needed to gen-erate a new life form. This also makes eggs one of the best possible food sources for other animals, as they're nutrient-dense with no ability to run away or fight back. Humans figured this out pretty quickly, and we now farm chickens to produce more than three bil-lion eggs per *day*.[3] Although we're the only species that has indus-trialized our egg consumption, we're far from the only species obsessed with eating eggs. Foxes, weasels, rats, dogs, snakes, crows, and many others are all avid egg eaters. Fish eagerly slurp up the eggs of other fish, or even their own eggs if they're low on energy.

Because of the inevitable loss to predators, animals produce more eggs than are needed to sustain the population. This eggy excess injects nutrition into countless food webs, connecting diverse species and ecosystems. Land-dwelling raccoons scarf down the eggs of seafaring turtles; tree-climbing snakes swallow the eggs of sky-faring birds. The feedback loop in which eggs get eaten, so parents make extra eggs, so more eggs are available to be eaten, means that animal babies matter to the world—as a rich food source—long before they even hatch.

But safety in numbers isn't an egg's only hope of survival. Animals have also evolved numerous clever strategies to hide and protect their eggs: shells, capsules, nests, pouches, wombs. It's important to note that these protections fail if they isolate the embryos too thor-oughly. An obvious example is that an egg completely impermeable to gas will prevent the baby from breathing! Eggs cannot be impenetrable walls that separate the developing baby from the environment but must instead filter and interpret that environment for the baby. From its very first cellular division, an embryo's trajectory is shaped by the world around it. The nutrients placed into the egg by the mother are determined by her diet and overall health. Oxygen, temperature, and chemicals all cross eggshells and membranes, as do the physical vibrations of a hunting predator.

This, then, is "what's so special" about a fertilized egg cell: It is a package of raw materials, instructions, and machinery, uniquely capable of growing into an animal, and preloaded with environmental input. Signals of available food and potential danger are already integrated into the egg cell by the time development is kicked off, a moment that is itself determined not by some internal genetic clock but by an external encounter with a sperm cell. Eggs can lie dormant for decades before fertilization sparks them to action. In this chapter, we'll follow the course of embryonic development as a cell becomes a baby, shaped by its environment every step of the way.

Everything that's inside an egg (and a sperm)

Why can't we grow a whole new animal from a blood cell or a brain cell or a skin cell? People have wondered about this for a long time. One old idea was that sperm contained the raw baby-making material and it only needed "planting" in the "fertile soil" of a uterus. Hence the word *semen*, which means "seed." We now know that eggs and sperm are both special, each containing only half as much DNA as other types of cell. At fertilization, these two halves combine to form a new whole. While egg and sperm contribute nearly equal amounts of DNA, the egg contains additional necessities: yolk to feed the growing embryo, proteins to decode genes, mitochondria to provide energy.

Mitochondria may not sound as familiar as yolk or protein. However, your body contains more mitochondria than cells!

They are our energy factories, with some cells containing more than a thousand mitochondria each. Animals didn't invent mitochondria, though—we co-opted them. Mitochondria are the descendants of free-living bacteria that were engulfed, more than a billion years ago, by the precursors of animal cells. Over time, these bacteria became simplified into part of our cellular architecture. However, they kept their own DNA, isolated inside themselves from the rest of our DNA, which is stored in the cell's nucleus.

Both egg and sperm cells contain mitochondria, and, in fact, mitochondrial energy is crucial to the swimming of animal sperm. However, in the vast majority of animals, sperm mitochondria are destroyed after fertilization.[4] (Mussels are a peculiar exception: Male embryos preserve their father's sperm mitochondria and sequester it in the cells that will eventually produce sperm, so male mitochondrial DNA is passed from sperm to sperm across the generations. No one knows why.[5])

Thus, while you inherit nuclear DNA from both parents, you inherit mitochondrial DNA only from your mother. Scientists have even shown that female mice can filter out badly mutated mitochondria when they're making egg cells, so their offspring receive the best of the best.[6] It seems likely that other animals, including humans, could do the same. As explained in the wonderful tune "A Biologist's Mother's Day Song," the combination of nuclear and mitochondrial DNA along with nutrition and proteins means that "slightly more than half of everything I am is thanks to you [Mom]."[7]

In many species, eggs are so self-sufficient that they can develop into fully functional adults without needing sperm at all. Unfertilized honeybee eggs develop into adult males; unfertilized aphid eggs develop into adult females. Unfertilized eggs of many lizard species develop into adult females, and this system works so well that at least one species has no males left whatsoever. Some unfertilized bird eggs have been found developing into embryos, but they usually die in the shell. A striking exception was reported in 2021, when scientists at the San Diego Zoo's condor breeding program realized that two unfertilized condor eggs had successfully hatched.[8] This kind of reproduction without fertilization is called *parthenogenesis*, and although it's central to a few species' reproductive habits, across the animal kingdom it is the exception rather than the rule.

Genetic contribution from a sperm cell is usually required for development to begin. That's the reason the chicken eggs in my fridge and the human eggs in my ovaries aren't busy growing into animals, while the songbird eggs in the nest and the spider eggs under the rocks certainly are. Recent research is even uncovering nongenetic sperm contribution in some species. Although there's definitely no tiny person inside a sperm cell, fully formed and ready to unfold as the "preformationists" once believed, there *is* something that I find nearly as astonishing: information about environmental conditions.[9] Fathers who experience a deficient or unbalanced diet, or exposure to toxic chemicals, change their sperm in ways that are passed along to their offspring—possibly because such changes proved adaptive to a difficult or dangerous environment in the past.

The magic of making embryos from scratch

In 2008, two years after the discovery of the Humboldt squid egg mass in the Gulf of California, I took a couple of big steps forward on my path of understanding reproductive biology. First, I got married, and second—following a long-standing post-marital tradition—I made a lot of babies. However, they weren't human babies, and I made them independently of my new spouse.

We were living in Monterey, California, both of us working at Hopkins Marine Station, a satellite campus of Stanford University and the oldest marine research laboratory on the West Coast of the

United States. I'd been a graduate student there since 2005, and my partner, Anton, had come to join me in 2007. He was an engineer, so the biomechanics lab had eagerly hired him to build their specialty equipment.

I had recorded the behavior of the squid hatchlings from the egg mass, along with all my attempts to feed them, until the last one perished a couple of weeks after our return to California. No more egg masses had since been found. (Another Humboldt squid egg mass wouldn't be discovered by scientists until 2015.[10]) I'd continued using *in vitro* fertilization to get the babies I wanted to study, but it was a struggle to produce live, healthy embryos. I needed more training.

I set my sights on a world-renowned developmental biology class at Friday Harbor Marine Laboratory—the *second* oldest marine station on the West Coast. Every summer for a hundred years, marine biology students and researchers from around the world have converged at Friday Harbor in Puget Sound, off the coast of Washington State, to collaborate, teach, and study.

I applied and was accepted to the class, which commenced two weeks after my wedding. Anton and I decided on an unconventional honeymoon, driving up the coast from Monterey to the San Juan Islands of Puget Sound. After several damp campsites, one tick bite, a party at the new in-laws', and my first ride on a car ferry, I parked my dusty old stick shift in the Friday Harbor lot. Anton flew back to Monterey, and I proceeded to spend the next month doing what new brides throughout history have habitually done: spawn.

Over the weeks at Friday Harbor, I created baby starfish, sea urchins, sea cucumbers, sea snails, sea slugs, acorn worms, bristle worms, shrimp, crabs, and moss animals. I was introduced to a larva that looks like a tiny hippopotamus, and another that eats its own body like an ouroboros. I learned about the tremendous range of reproductive habits in the sea: Some species release eggs and sperm freely into the water, leaving fertilization up to the cells themselves. Others fertilize their own eggs but then let the embryos loose. Still others guard their fertilized eggs on the seafloor, or even incubate them inside their bodies until hatching.

From the very first day, the importance of cleanliness in an embryo laboratory was impressed upon us. All the glassware, the dishes and beakers and cups that we used, were marked "E" for "embryo," to indicate they had never been touched by preservatives, had never been cleaned with anything but water. At the time, I saw this as a necessary step toward our end goal: successfully developing animals. Reflecting on it now, I'm fascinated by the implied sensitivity of embryos. What does it mean for animal development in the wild, that pollutants of many kinds have become near ubiquitous in the water, the air, and the soil? Is anything in the real world "embryo-clean" anymore? Perhaps not, and yet development has not ground to a total halt. Undoubtedly there are many hidden changes we have not yet quantified: reductions in the likelihood of survival to hatching or metamorphosis, adaptations by parents to increase the protection of their eggs or by embryos to cope with an impure environment.

Our study of development began with fertilization, as we learned how to extract eggs and sperm from various adult animals and mix them in appropriate ratios. We reveled in a mystery that has fascinated biologists since they first began observing fertilization under the microscope: Despite the thousands or millions of sperm outnumbering each egg, only a single sperm nucleus enters it. Many early embryologists held on to the old notion that sperm contained all the material that mattered in development, so they assumed that the sperm itself must be responsible for preventing multiple entry. However, in the first decade of the 1900s, the great developmental biologist Ernest Everett Just (1883–1941) showed that the incredible architecture of the egg cell prevents fertilization by multiple sperm. In fact, he observed two distinct sperm-repellent changes in a fertilized egg: a fast electrical block and a slower mechanical block.[11] (Scientists have now found two exceptional groups of animals that *do* permit multiple sperm inside the egg: comb jellies and arrow worms. In a phenomenon that still boggles my mind whenever I think about it, the nucleus of these egg cells *chooses* one of the sperm nuclei to merge with, forming the genome of the new individual. The remaining sperm nuclei disintegrate.)

Just went on to elucidate how the cells of the newly formed embryo are able to continue adhering to one other as they divide. Alongside his focus on the cellular aspects of development, he retained a passionate regard for the importance of the natural environment. He wrote, "In a certain sense we should not speak of the fitness of the environment or the fitness of the organism: rather, we should regard organism and environment as one reacting system."[12] But even as Just was making a clear case for the unity of environment and development, the field was in the process of rupturing into two domains: embryology and developmental biology.

DNA and its famous double helix had not yet been discovered. Scientists were still searching for "determinants," or instructions that tell an embryo how to grow. Thomas Hunt Morgan (1866–1945), who began his career as an embryologist, founded what we now call *genetics* when he identified the location of these determinants. He conducted a series of experiments to prove that chromosomes (the structures inside cell nuclei, which we now know are formed of DNA strands like intricately spooled-up thread) contain the mechanisms of heredity, and that

Ernest Everett Just earned a doctorate in zoology from the University of Chicago, but racism in American academia curtailed his professional opportunities. Many of his brilliant experiments were conducted in the more welcoming European environment.

chromosomal changes lead to mutations. This opened up the possibility of figuring out which chromosomes—and eventually, which DNA sequences—controlled which bits of development. Studies like these looked at the organism as a self-contained system and led to a major shift in thinking, away from ecological studies of animals in their environments and toward abstract, idealized models.[13]

Thus developmental biology gave rise to, and remained entangled with, genetics. No longer interested in watching embryos develop in natural conditions, developmental biologists moved indoors, building labs in large universities and medical centers. Meanwhile, the embryological tradition kept its head above water at marine stations around the world, where sea urchins, sea snails, and other abundant animals provided an endless supply of eggs.[14] Friday Harbor was one such bastion of classical embryology. I was fortunate to not only spend a summer there but also to learn from an embryologist among embryologists, the inimitable Richard Strathmann.

Richard must be described in conjunction with Megumi Strathmann, as the two are partners in both marriage and professional life. They are embryo experts, as kind and generous as they are intellectually curious and rigorous. The most notable difference between them is their height. (This may have predisposed me toward a fondness for the couple, as I, like Megumi, am a full head shorter than my mate.) They arrived in Friday Harbor as newlyweds in 1965, and over the decades since, the Strathmanns have conducted such elegant research and taught such illuminating classes that nearly everyone I interviewed for this book had a Strathmann connection.

What I remember best from that summer in Friday Harbor is a constant sense of wonder. Every day, every hour, our embryos showed us something new. One cell became two, then four, then eight. An undifferentiated ball became a body, with a front and a back, an inside and an outside. Embryonic blobs with no obvious similarity to any particular animal suddenly sprouted distinctive shells and skeletons, arms and tentacles.

Every cell in your body (except your symbionts, which we'll discuss in more depth later) is a product of that one fertilized egg. Every cell has the same genes. The reason that your skin cells look like skin, and not like a liver, is that different genes are turned on and off in different cells over the course of development. Before all that specialization, when you're only a few cells, each cell has the capacity to do almost anything. Damage to the embryo in this very early stage is often easily compensated for because the cells are so flexible and powerful.

Early embryos possess the astonishing ability to heal punctures and tears. This fact was crucial to embryological research from the late 1800s onward, since scientists were constantly cutting, poking, and injecting embryos to figure out how they worked. But it wasn't until the 1990s that people began to investigate the repair process itself. Then, several studies found that an embryo can either stitch itself back together with small pieces of membrane or apply a large membrane patch if the damage is severe. Given that most embryos will never be sliced open by a scientist's scalpel, this may seem like an unnecessary talent, but the natural world is no kinder than the laboratory. Repair skills would come in handy after damage from rough waves, storms, or predators.[15]

Normally, as Just found, embryonic cells stick together while they multiply, eventually forming rudimentary organs and systems. But if the adhesion between cells is broken and the embryo separates into two or more pieces, each piece can often develop independently into a functional adult. In fact, that's where identical twins come from. Although uncommon in humans, this kind of twinning occurs with great regularity in other species. Every fertilized armadillo egg divides into four separate embryos, each with exactly the same genes. The mother armadillo proceeds to gestate and give birth to identical quadruplets.[16] And if four seems like a lot, wait until you meet the parasitic wasps.

Wasps: terrors, tools, or just good moms?

One of my only memories from first grade is the yellow jacket wasp that got stuck in my school uniform and repeatedly stung the back of my knee. I know I'm not alone in my subsequent aversion to these black-and-yellow insects—fear of wasps is even more prevalent among humans than fear of bees.[17] However, when I learned about the habits of *parasitoid* wasps, the idea of a mere sting faded to an almost laughable threat.

Assuming that you are human and not an insect yourself, I encourage you to continue reading for a shot of schadenfreude, as you can rest assured that parasitoid wasps won't be targeting *you*. Most insects, by contrast, exist in a state of constant vulnerability to one or more parasitoids. These wasps don't look too different from yellow jackets, although they are often much smaller, colored a simple brown or black, and in place of a stinger possess an elongated ovipositor, or egg-laying device. This is used to inject their own babies into the eggs or larvae of other insects. The tiny wasp larvae grow up inside their living home, snacking on the larger larva's innards, until they're ready to metamorphose into adults. Then they eat their way out.

Some wasp moms deliver a paralyzing sting that renders their prey immobile, but still alive, while the wasp baby munches away. There's a good reason for this nightmare fuel. A dead body would be quickly taken over by microbes that make it unpalatable or downright toxic for wasp babies to eat. A live body remains fresh and nutritious.

Many other parasitoid wasps don't even need to paralyze their prey. The development of their young is exquisitely timed so that the host larva can continue to go about its business, eating and growing as it is devoured from the inside. It only dies when the wasps that it's been unwittingly incubating burst free—and some hosts even survive a little past that grisly experience. None, however, make it to maturity and reproduction of their own. This is the reason we call these wasps *parasitoids*, animals that always kill their hosts, rather than *parasites*, animals that rarely kill their hosts. (Proper parasites, from tapeworms to viruses, are most successful when they keep their hosts alive and actively spreading the parasites' offspring to new hosts.)

The parasitoid strategy works so well that it may have made wasps the most species-rich group of insects on the planet, perhaps even outdoing the famously diverse beetles—which, at several hundred thousand species, constitute a quarter of all described animal species.[18] (It's thought that most wasps have not yet been described.) Parasitoid wasps have been able to diversify into so many species because they can specialize, and one of the most remarkable specializations occurs in the subgroup of parasitoid wasps that lay cloning eggs. This goes far beyond twinning or quadrupletting—a single wasp embryo multiplies to become *thousands*. The resulting siblings are genetically identical yet have the capacity to develop into distinct larval castes like those of an ant colony, with "soldier" larvae that die without pupating and "reproductive" larvae that can mature to adulthood. Their differentiated development must be driven by environmental rather than genetic cues and has been best studied in the wasp *Copidosoma floridanum*, which is tinier than a gnat and lays its minute egg in a moth caterpillar. Scientists have found that a cloning embryo of this species produces more soldier larvae relative to reproductive larvae when it detects the presence of a different wasp species parasitizing the same host. This extra investment in defense (or rather, offense, since the soldier larvae seek out and kill unrelated wasp babies) is only worthwhile when there's something to defend against.[19]

In other species, it's not the embryo but the mother wasp who assesses the environment and makes decisions based on what she finds. A female of the also less-than-gnat-size species *Nasonia vitripennis* examines a potential host looking for eggs laid by other members of her own species. This information determines whether or not she will use stored sperm (which she keeps in a receptacle after mating) to fertilize her eggs. Like honeybees, wasps produce male offspring from unfertilized eggs and female offspring from fertilized eggs, and the wasp mom benefits from producing more of whichever sex is likely to give her more grandchildren. Daughters are reliable producers of grandchildren, since they don't even need to mate to lay viable eggs. If a son can't find any mates, he produces no grandchildren; on the other hand, if he finds many mates, he can father more grandchildren than his sisters. Thus, a mother wasp adjusts her eggs' sex ratio based on how many other wasps' eggs are already inside the host insect that she plans to use as a nursery. The more other eggs, the more likely there will be available females for her sons to mate with, and the more unfertilized eggs she lays.[20]

Although we might be inclined to think of a wasp that fills another insect's baby with her own babies as a "bad guy" (indeed, Charles Darwin considered them a puzzle to challenge the existence of a Creator both omnipotent and benevolent), we humans actually use these wasps as "good guys." *Copidosoma* goes after moths that devour crops; *Nasonia* targets disease-carrying flies. Other species of parasitoid wasps have been successfully released as biological controls to halt invasions of agricultural pest species, and the better we understand the reproduction and development of these strange insects, the more effective our deployment of them can be.[21]

We are a group of collaborative cells

"Forming an embryo is the hardest thing you will ever do," begins my college *Developmental Biology* textbook. "You had to build yourself from a single cell. You had to respire before you had lungs, digest before you had a gut, build bones when you were pulpy, and form orderly arrays of neurons before you knew how to think. One of the critical differences between you and a machine is that a machine is never required to function until after it is built. Every animal has to function as it builds itself."[22]

Put that way, it sounds so difficult that one has to wonder why we do it at all. Although it may seem like we don't have a choice, other kinds of reproduction are possible. An adult sea anemone can clone itself, splitting in two. So can corals, sponges, some starfish, and many worms. In general, an adult that's ready to reproduce by making eggs and/or sperm is a successful adult. It's healthy, and it's probably in a good location. Why not replicate to create two successful adults—why generate single reproductive cells that have to build a whole new body from scratch?[23]

Sex is one very good reason. Cloning or budding, anemone style, is a form of asexual reproduction, which can only create new individuals that are genetically identical to the old ones. Sexual reproduction is thought to confer a major advantage because it allows genes to mix and shuffle, creating genetically novel individuals that may be better suited to a changing and variable environment. Simplifying to a single cell before you mix your genes with those of another individual means that the mixed genes will be consistently represented throughout the resulting organism.

Another good reason is to shake off pathogens. Very few diseases or infections can be transmitted from a parent's body into an egg or sperm cell. Worms inhabiting the gut, mites crawling on the skin, parasites traveling with the blood—all of these are shed in the transition to a single cell, which has neither gut, nor skin, nor blood.

A third reason, unproven and provocative, has to do with competition and cooperation between cells. All your billions of cells contain the same genes, since they're all derived from the same fertilized egg. But over the course of your lifetime, they've had different experiences. Different genes have been turned off and on, and different hormones and chemicals have surrounded them in different concentrations. This buildup of variation introduces the possibility that cells belonging to the same organism may experience different costs and benefits to cooperation. Starting each new organism from a single cell "resets" the system, wiping away accumulated differences between cells that could lead to conflict. All the cells in the new organism once again begin with the same environmental history, possibly making them more capable of collaborating to produce a functional animal.[24]

Cancers are dramatic examples of breakdowns in cellular cooperation. The cells in a tumor are "defectors," proliferating and appropriating resources to benefit themselves while damaging or dooming the rest of the organism. However, they can very rarely be passed from parent to child, as animals set aside the cells that produce eggs and sperm early in development, sequestering them from the rest of the body. Thus the "single-cell bottleneck" may have evolved to protect future generations from renegade cells, ensuring a cooperative restart with a fresh slate.

"We are a group of collaborative cells, walking around," in the words of biologist and sculptor Fernanda Oyarzun, who as Strathmann's graduate student helped teach my embryology class in Friday Harbor. In addition to her talent for research and instruction, she shared her artistic skills, encouraging us to use the act of drawing embryos to observe things that might be less visible in a photograph. Oyarzun's wonder at developmental phenomena was contagious. "Looking at an organism developing under the microscope, it's like seeing a sculpture making itself," she says. "I couldn't believe a cell would just go, boop, and then it's two cells, then four. It's crazy."[25]

At this point, I must regretfully announce that embryologists' utter fascination with early development led them to invent a *lot* of new terminology to describe it. Like all biological jargon, it has its uses and becomes somewhat easier to grasp when we can visualize it. The very earliest stage of an embryo is called a *zygote*—that's the thing that goes "boop" to double its cells, then double them again. Eventually it becomes a ball of cells called a *morula*. Morula cells

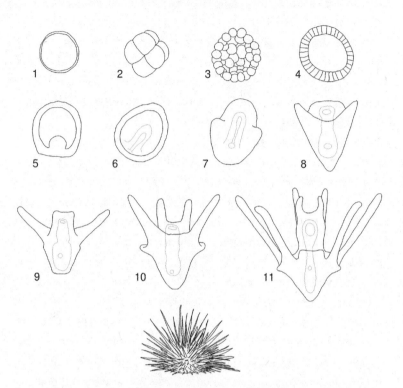

Stages of sea urchin development: 1) fertilized egg; 2) eight-cell stage; 3) solid ball of cells, or morula; 4) hollow ball of cells, or blastula; 5) beginning of a gut, or early gastrula; 6) gastrula with complete gut; 7) prism; 8) early pluteus; 9) two-arm pluteus; 10) four- to six-arm pluteus; 11) eight-arm pluteus. Not to scale.

are still pretty much all the same, and the morula is the stage at which those wasp embryos divide into so many identical siblings.

Next, the cells of a morula migrate toward the periphery, forming a hollow ball called a *blastula*. The process is similar to a dense cluster of children being coaxed into a circle to play a game, except it happens in three dimensions. (Imagine children floating through the air and holding hands to produce a sphere rather than a ring.) All animals go through a blastula stage, although fruit flies and other insects engage in a curious modification. As a zygote, a fruit fly multiplies its cell nuclei but not its cell walls. Thus, as a morula, a fruit fly is essentially one enormous cell containing hundreds of nuclei. To form a blastula, these nuclei migrate to the edges of the cell, just like individual cells would migrate in other animals, at which point the fruit fly finally grows cell walls between its nuclei and begins to look more or less like the blastula of other species.

You might expect that the hollow interior of a blastula would become the animal's gut. But no. Our guts are technically "outside" our bodies, formed in one continuous surface with our skin. This is how that process begins: A dent appears in the blastula, as if a tiny finger had poked it. This dent will become either mouth or anus, and it grows deeper and deeper until it reaches the other side of the blastula. There, the remaining requisite orifice, anus or mouth, will open up. The embryo is now called a *gastrula*, from the ancient Greek word for "gut."

Whether the gut begins growing from mouth or anus provides one of the most fundamental divisions in the animal kingdom. You could have a table covered with blastulas from dozens of different species, and they would all be more or less the same until the moment of gastrulation. At this point you could draw a line down the center of the table and divide the embryos. On one side go the *protostomes*, which means "mouth-first." This includes the snails (a group far bigger than garden snails, encompassing nearly every seashell you can imagine) and all their relatives, from clams to cuttlefish, as well as every insect, spider, crab, and worm. On the other side go the *deuterostomes*, a word delicately chosen because it means "mouth-second" rather than "anus-first." This group includes the echinoderms (sea urchins, sea cucumbers, and sea stars) as well as every vertebrate—sharks and sheep and gulls and gorillas. And, of course, humans. Calling someone a deuterostome has become a favorite insult in my family, along the lines of "your epidermis is showing." Our deuterostome kinship is one reason we've been able to learn so much about our own development from studying sea urchin embryos and larvae. For example, it was research on starfish larvae in 1883 that revealed the existence of specialized defense cells.[26] Scientists proceeded to find similar cells in other animals, including humans, kicking off the whole science of immunology.

At the gastrula stage, certain species of fish can enter a kind of suspended animation called *diapause*. These "annual fishes" live in temporary pools that appear only briefly after rains, so they spend most of their lives in diapause. Flexibly adjusting their development to a variable environment is so important that they've also evolved the ability to enter diapause at two distinct times later in embryonic development. When rain falls and puddles form, the eggs promptly hatch, grow up, reproduce, and bury more eggs in the mud. Adults die when their puddles dry up, but if they're lucky they've already seeded the next generation.[27] Diapause isn't unique to fish; several kinds of shrimp create "resting eggs" that can be buried in sediment, only to hatch out later and resume an aquatic way of life. Free-living roundworms can create an extra developmental stage called a *dauer* if there isn't enough food in their environment. Dauer larvae exist in stasis, not eating for months, until conditions are welcoming enough for them to molt and resume their developmental march to adulthood. This ability may have facilitated the evolution of parasitic roundworms (which you may be familiar with if you've ever had a puppy), as a dauer larva could remain infectious longer than ordinary larvae, giving it more time to reach a suitable host. Even many mammal embryos can put their development on hold until conditions for gestation improve. We're still discovering the enormous amount of resilience built into embryonic development.

We're also just beginning to understand how often it fails.

Parrot eggs and hidden deaths

Biologist Nicola Hemmings's most surprising discovery began with a seemingly unrelated interest in sperm competition. She shared Just's curiosity: Of all the sperm trying to get into an egg, which one makes it? But instead of focusing on the cellular mechanisms of sperm blockage, she was interested in ways that females might influence which sperm was successful.

In graduate school, Hemmings's supervisor encouraged her to study not only contemporary research but also older journals that might hide forgotten clues and techniques. Thus began her exploration of the history of poultry science. Because of humanity's ravenous hunger for chicken eggs (as well as chicken meat), there's a lot of money in studying poultry, and scientists have invested a great deal of time in understanding fertilization in chicken eggs. Hemmings found these old papers "really badly written and boring" but still immensely worthwhile. Poultry researchers had figured out how to peel off an egg's membrane and look at it under the microscope to find out if sperm had penetrated. Hemmings built on this technique, adding fluorescent dyes to visualize sperm, and then she realized she could use the same fluorescent dyes to look for embryonic cells.[28]

For many years, eggs that failed to develop to hatching had been considered infertile. Researchers assumed that sperm had never successfully reached these eggs. Now, not only could Hemmings definitively identify whether sperm had entered, she could say for sure whether or not embryonic development had begun. "Have we been completely thinking about this the wrong way round?" she wondered. The answer was yes. According to her experiments, it was far more common to find that fertilization had been successful, but embryos had died in stages too early to observe with the naked eye.[29]

As her research expanded beyond sperm competition into embryo mortality, Hemmings began working with bird conservation programs to elucidate reproductive problems, a necessary precursor to figuring out practical interventions. Her program now includes birds from threatened populations around the globe, especially New Zealand, where birds evolved in the complete absence of mammalian predators. Before humans arrived, the greatest dangers to their eggs and chicks were other birds, particularly the

The total population of kakapo had dropped to only 51 by 1995, but protection from predators and managed breeding have brought that number up to 252.

sharp-eyed Haast's eagle, so they adapted nests and burrows hidden near the ground. Unfortunately, such habits offer no protection at all against the ferrets, foxes, cats, and rats that European colonizers brought to New Zealand.

The kakapo, the world's only flightless parrot, is now critically endangered. The survivors have all been moved to predator-free islands, where they are intensively monitored. In 2019, a kakapo scientist contacted Hemmings to ask her to analyze all the eggs that were presumed infertile. Kakapo breeding seasons are irregular and usually result in only a handful of eggs, with about 65 percent failing to hatch. This had led to an estimate that about half the living males were infertile. That year, however, the kakapo produced a record number of eggs, and those that didn't hatch got shipped to Hemmings for analysis.[30] She and her colleagues discovered that the New Zealand parrots had no major issues with fertility; instead, many of the embryos were dying early in the egg.

Because kakapos are tracked as carefully as celebrities, the population managers have records showing which mating events led to which eggs. Now, instead of writing off unsuccessful fathers as infertile, they can look at each mating pair as more or less likely to succeed. If a certain combination of male and female always seems to produce embryos that die early, these individuals can be separated and moved into different populations to promote matches with a higher likelihood of hatching success.[31]

In addition to the importance of early embryo mortality for conservation, Hemmings notes that it can have a striking influence on studies of evolution. In another New Zealand species, the heihei, male embryos are more likely than female embryos to suffer early death. If scientists had never found that difference, they might have noticed a biased sex ratio in chicks and adults, then looked for all kinds of explanations that have nothing to do with what's actually going on. (To be clear, we still don't know what's actually going on.) That's not only the case for sex ratios. Most traits, from color to size to speed, are studied by sampling populations long after the early embryo stage. But if any of these traits are linked to non-random early death, then sampling from the adult or even the juvenile population is missing a big piece of the picture. A true understanding of trait distribution would need to include all the individuals that died before samples were taken.

Early deaths only detectable by microscope are likely caused by genetic failures. Embryos that survive this stage to continue their development go on to face two more significant killers, more visible but no less destructive. One, of course, is predation. Eggs are a tasty treat, full of nutrition, unable to run away—right? In fact, egg predation is *such* an enormous risk that it has created enough selective pressure for tree frog embryos to genuinely "run away," by hatching early. When a hungry snake begins to chow down on a batch of frog eggs, the ones that aren't immediately consumed can hatch up to two days before they would have otherwise, swimming away to escape their siblings' fate.[32] A contrasting avoidance strategy has evolved in bamboo shark embryos. When they sense an approaching threat, these babies freeze and hold their breath in hopes that the hungry predator won't notice them.[33] Thus, even within the egg, an embryo can improve its odds of survival by perceiving and responding to its broader environment.

The final source of in-egg mortality, which surprised me to learn about, is the process of hatching itself. "The egg's brilliant, it does everything you need, but it's a hard shell, and you've got to crack out of it," Hemmings explains. "As chicks get bigger, everything has to be perfect. If you're not in the right position, if you haven't got enough energy, if you've run out of water or whatever, then that's not going to happen." Eggshells are so tough that embryonic birds and reptiles have had to evolve special "egg teeth" to break free. Even if their parents incubate them before birth and provide food afterward, hatching is something all these babies have to manage on their own. Only crocodiles have a mother that will step in (or rather, bite in) to help them out of their shells.

Embryos embody a curious contradiction: They are at once our most resilient life stage and our most vulnerable. Luckily, embryos aren't in it alone. Their parents have a vested interest in making sure they get through this stage and on to the next. How do parents ensure that the embryo gets everything it needs—nutrition, water, temperature, microbes and defenses—to complete its astonishing bodybuilding job, to avoid being eaten by predators, and to escape successfully from its protective prison? That's what we'll explore in the next three chapters: How animal babies benefit from their parents' provisioning, brooding, and gestating.

2

PROVISIONING

From Edible Siblings to Algal Life-Support

What did I know, what did I know
Of love's austere and lonely offices?

—Robert Hayden, "Those Winter Sundays"[1]

Ignoring outliers like ducklings, kangaroos, and humans, the vast majority of animals begin their lives without a parent nearby. Squid die after spawning, their bodies long since devoured or decomposed by the time their eggs hatch. Sea turtles swim hundreds of miles out to sea before the babies they've buried in sand come to light. Corals, sea urchins, and mussels all fling their naked eggs to the wide waters. These "orphan embryos," as scientists call them, must fend for themselves even as blastulas. But don't be fooled—their parents are no less invested in the success of their offspring than the elephant who nurses her baby for years. Their investment simply looks different.

Every parent aims to produce a child that can thrive wherever it ends up and, at the same time, puts some effort into curating where the child ends up. So parents face two interrelated tasks: first, to make a baby that's big and strong and capable, and second, to ensure said baby arrives at an excellent place to live.

This is the parents' chance to travel into the future. Very few (very, *very* few) animals have any hope of living forever. (Immortal jellyfish and four-hundred-year-old sharks come to mind as exceptions, which we'll discuss in the Epilogue.) Throughout history, plenty of humans have compared having children to achieving a kind of immortality—while plenty of children have resented representing such a thing to their parents. As the field of genetics took off, reproduction took on the definition of "immortality for our genes." But in recent years our understanding of genetics has been changing. We see now that genes do not remain fixed throughout an organism's life and can be altered by environment and experience. Genes are far from inert time capsules passed from body to body.

The dinosaur ancestors of modern birds would not look at their great-great-great-etc.-grandchicks and think "Ah, immortality!" Having children is less of a chance to live forever and more of an opportunity to offer future generations the resources and experiences we've gathered from our time in the world. Our offspring and grand-offspring will use what we give them, in collaboration with the world around them, to mold something new.

Of course, if babies don't survive, there are no future generations. Survival, therefore, is the number one goal. Evolution selects for parenting strategies that promote the survival of offspring. And each strategy has to interact successfully with the environment—an egg built to keep an embryo moist on land will suffocate the embryo if placed underwater, while an egg built to keep an embryo well-oxygenated in the water will dry out in the air. Parents both protect their offspring from and introduce them to the environment.

How to pack your baby's lunch

Eggs can be as tiny as shrimp eggs, like specks of dust. They can grow as huge as ostrich eggs, the size of a human baby's head. (The ostrich egg is often called the largest single cell in the world; although some nerve cells can grow longer, none are nearly as heavy.) But that giant ostrich egg doesn't contain a giant embryo right at fertilization. It contains a small embryo, attached to an enormous yolk. Yolk is the delicious fatty goo that Mom packs inside the egg to feed her baby—and she plays favorites.

Bird parents have evolved techniques to feed some embryos more than others. Female birds often mate with numerous males, judging each partner by species-specific criteria of attractiveness and success. When they lay their eggs, they offer an extra boost to those fertilized by what they deem to be the best fathers: increased egg size, and therefore yolk content, which leads to larger and more competitive chicks.[2] Blue-footed booby females assess the blueness of their mate's feet as a signal of how well fed he is, and if scientists artificially color a male's feet darker (less well fed) then his partner will lay smaller eggs, reducing her own investment in the children of this clearly inferior mate.[3] Mallard moms will similarly lay smaller eggs when dad is less attractive, although scientists still aren't sure how to measure male mallard attractiveness and must rely on female mallards to show them who's hot and who's not.

Not all animals can adjust the size and yolk content of their eggs as much as birds can. Many egg-layers employ a different tactic for getting more resources to the offspring who are likely to do the best. They begin by packing all the siblings into a single capsule, along with extra snack eggs. What better nutrition for your babies? Moms often don't bother to fertilize these "nurse eggs." Some developing embryos swallow their nurse eggs whole; others wait for the eggs to break down, then collect and eat the particles. Some sea snail embryos start to eat their nurse eggs when they're barely more than a gastrula themselves, as soon as they have a mouth. They can't digest the food right away; they just pack it in a ballooned-out pouch. It's almost like they're using the nurse eggs to create their own internal yolk sac, which they'll live off as they develop.[4] The most competitive embryos get the most nurse eggs, without the mother having to do any work to figure out which are the most competitive. She simply puts them all together and lets them shake it out.

Competing with your siblings for food while you're still a developing embryo sounds stressful enough, but many species take sibling rivalry even further. In these species, *all* the eggs are fertilized, and all have the capacity to develop to hatching. They also have the capacity to eat each other. These babies find themselves enmeshed in predator-prey dynamics even before they emerge into the wider world. From the mother's point of view, it's just another way to ensure that the babies who come out are the strongest, most competitive, and best fed.[5] From biologists' point of view, it's another opportunity to coin a new word: *adelphophagy*, which means brother-eating. (One of my professors offered a truly incredible mnemonic for this bit of vocabulary. "Philadelphia is the city of brotherly love; Adelphophagia is the city of brotherly cannibalism.")

A model system for studying adelphophagy can be found in the wonderful group of animals known as bristle worms. They are related to earthworms but look like surrealist fever dreams by comparison. Bristle worms can be as iridescent as peacocks or as luminous as fireflies, with tentacles resembling a pile of spaghetti or a lacy Christmas tree. They can burrow in mud, eat through rock, or build their own reefs. Some sting so fiercely they've earned the name fireworms, while others are harvested as a delicacy worthy of feasting. Among this wild diversity, bristle worms all share a few features: They are elongate, segmented, and produce a marvelous little larval form called a trochophore. We'll spend more time with trochophores later; for now, we'll focus on the parenting strategies that bristle worms have evolved to cope with a highly variable habitat.

Some worm parents carry egg capsules around on their backs, while others lay their capsules on the seafloor. They can load the capsules with various types of eggs to either facilitate or discourage sibling cannibalism, and rather than letting the babies hatch on their own schedule, the mother decides when to rip open the

In other species, an embryo's cannibalistic tendency is not encoded in what kind of egg was laid but is instead dependent on how closely related its siblings are. Females of many species mate with multiple males, so it's possible for one mother's offspring to be full siblings (if fertilized with sperm from the same male) or half-siblings (if fertilized with sperm from different males). The Pacific sea snail *Solenosteira macrospira* produces egg capsules containing both relationships, and evolutionary biologist Rick Grosberg figured out that half-siblings cannibalize each other much more than full siblings. "Promiscuous" mating systems have long been considered advantageous to females because they increase the genetic diversity of their offspring, but it's possible that there's also an advantage in driving up competitiveness between siblings. "Okay kids, I'm setting up the most competitive situation I can," Grosberg speaks in the hypothetical voice of the mother snail. "I want to know which one of you gladiators is the best, and if I want to find the best, I want to make you as variable as possible."[8]

Depending on the environment, competition among these snail embryos may be more or less desirable. When embryos don't eat each other, they don't grow as large—but far more of them successfully hatch. As long as they can find food to eat after hatching, it may be better to produce many small offspring rather than fewer large offspring. Grosberg explains how food availability changes over the reproductive season of the snail, due to the settlement and growth of young barnacles, called spat. (We'll find out what these barnacle babies were doing before settlement in chapter 5.) Early in the season, two different factors come together to reduce sibling cannibalism. Mating has only just begun, so females are likely to lay capsules full of eggs fertilized by only one male. At the same time, barnacle spat have just begun to settle and are still very tiny. The snail babies are disinclined to eat each other because they are full siblings, and there's plenty of tiny food for them to eat as soon as they hatch. They hatch at small sizes, and they're fine. Later in the season, mating has been going on long enough that the baby snails in the capsules likely have different fathers. They're more inclined to eat each other, making the ones who do survive nearly twice the size of early-season hatchlings. Grosberg speculates that this is adaptive to the environment these late-season babies encounter, where barnacles have been growing, too, and only large snail babies can consume them.[9]

Birds and mammals rarely, if ever, eat their siblings, but they do engage in siblicide. Some bird chicks throw their siblings out of the nest, and hyena twins are notorious for prompt post-partum siblicide. It's tempting to assign human motivations and emotions to these situations, especially when I read about sibling cannibalism right after mediating my children's squabbles. Watching my kids quarrel is one of my least favorite activities. It's hard to fathom how animal mothers can deliberately set up their children to kill and eat each other. But, of course, it's a different situation. Unlike snail children, human children benefit tremendously from learning how to resolve conflicts without violence. Friction with their siblings serves a different, yet still advantageous, purpose in their development.

Although competition may be rife in the natural world, we're uncovering more and more examples of cooperation alongside it—like the symbiotic microbes that shepherd babies into the world.

The circle of life and the cycle of poop

Scientists are constantly discovering more species and strains and categories of microbes, and more ways that they influence the world around them. We know the names of some microbes (I can't wait to introduce you to *Wolbachia*) and we recognize the impacts of other as-yet-nameless microbes. We interact with many microbes that are free-living, as well as all the microbes that form our *microbiome*, living in and on our bodies and helping us accomplish the daily tasks of life. As we'll see, the process of acquiring a microbiome is integral to our development. It could be argued that microbes are the most important developmental influence on any baby animal; they play the roles of nursemaids, bullies, playmates. And although most microbes don't leave fossils, we have circumstantial evidence to prove that they've been playing these roles for a very long time.

In the mid-Cretaceous, if a small dinosaur died, its carcass had a chance of being turned into baby beetle food by a collaboration between beetle parents and microbes. These were the first burying beetles, the earliest example of parental care preserved in the fossil record.[10] The beetles themselves fossilized; the microbes are presumed to have been present based on similarities between these fossils and modern burying beetles. Today, mother and father burying beetles work together to find, process, and inhume an animal carcass that will be eaten by their grubs. We humans wouldn't feed our children meat that had been sitting out for days, because putrefying microbes produce toxins to repel us so they can keep the meat for themselves. But burying beetles carry different, cooperative microbes. These microscopic partners outcompete the putrefying microbes, thereby preserving the carcass for larval consumption.[11]

Think of all the animal carcasses these tiny undertakers have processed and buried over the last hundred million years, carcasses that would otherwise have rotted in the open. Dung beetles perform a similar service, burying poop that would otherwise remain redolently on the ground. In addition to their admirable accomplishments as sanitation workers, dung beetles have become a model system for ecological developmental biology. To understand how that happened, I called up Armin Moczek, a dung beetle researcher and German biologist now at Indiana University.

At the beginning of our Zoom call, he steps out and returns with a sizable praying mantis resting calmly on his hand. He tells me it's an invasive species, and that he collects them every year in large numbers as a teaching activity with one of his classes. "Then I don't want to bring them back, because they're actually really destructive, but I also can't kill them, they're just too cool. So they live out their lives in my office." I ask if they reproduce. They do. "Then I have hundreds of tiny mantids running through my office, but I still can't let them go because they're invasive, so they usually just eat each other."[12] That's one efficient use of sibling cannibalism.

In his research as well as his teaching, Moczek has a knack for utilizing species and systems that are close at hand. When he first came to the US as an exchange student, he collected insects regularly. He came across a dung beetle that could live on both cow and horse dung. Most of us don't spend much time thinking about the nutritional content of poop, but Moczek was curious about the difference in quality between the pungent products of cow and horse digestion, and how it would affect the beetle grubs eating each type of food. With laboratory tests, he proved that a grub raised on cow dung needed to eat twice as much as a grub raised on horse dung to grow to the same size. "But what was the big surprise and completely unanticipated was that *Mom knows the difference*. She makes her cow dung brood balls twice as big as her horse dung brood balls." Like the bristle worm mothers who adjust their babies' hatching based on water temperature, these dung beetle mothers assess the nutritional profile of available baby food to create an optimal meal for their developing grubs. These are parents as mediators of the environment for their babies.

Moczek now rears five different species of dung beetle in his lab and tracks every detail of their life cycle. The mother starts by digging a hole, then rolls a ball of dung into the hole. "The brood balls are half to a third of the volume of a golf ball. For a beetle the size of a coffee bean, that's a significant chunk of work," he explains. Inside the brood ball, the mother beetle leaves a tiny fecal deposit of her own called a pedestal, which is packed with the gut microbes she uses to digest her own meals. On top of the pedestal, she lays an egg, and then she seals up the brood ball. The grub hatches after a couple of days, eats the pedestal, then starts in on the rest of the poop. It produces feces of its own as it eats, and it works them into the brood ball. This mixes the microbiome from inside the grub's gut with the uneaten food outside its gut, "predigesting" the remainder of the brood ball.

A dung beetle's childhood takes place in a carefully curated environment of mammal poop, mama poop, and microbes. Poop is serious business to beetles, but to humans, it's ripe for comedy, making a popular subject for cartoonists such as Rob Lang of Underdone Comics.

Not only is the system teeming with microbes, there are other symbionts at play, too. The term *symbiont* and its synonym *symbiote* apply to an enormous range of interactions between species, as they literally just mean "living together." Additional terminology can be helpful for specifying the nature of the cohabitation: Parasites harm, mutualists help, and commensals are neutral roommates. However, relationships can shift and encompass both beneficial and detrimental aspects, so it's not always easy to pick the right label. (Surely this is why Marvel Comics' Venom is labeled a *symbiote*, its authors wisely choosing the most general term in order to cover the character's transition from villain to antihero.)

Roundworms, similarly, can act as both parasites and mutualists. These smooth, slender creatures that wriggle and flip like cracking whips are usually associated with disease, since certain species can cause illness in humans, livestock, and crops. However, other roundworm species actually help baby dung beetles grow. Adult beetles can carry a variety of roundworms, some of which are sexually transmitted from male to female and then deposited in brood balls along with the egg. A 2018 study showed that these roundworms interact with the beetle's microbiome, promoting bacteria that help with digestion and defending against fungal infections.[13] It's a wild scene, and like many discoveries of symbiosis, it's probably just the tip of the iceberg. Roundworms have been found in association with all kinds of developing insects, including termites and bees as well as beetles. The dung beetle study was simply the first to put them under the microscope and reveal their importance.[14]

It may seem like an idyllic youth for the grubs—supported by microbes and worms, sheltered underground from predators and changing temperatures. However, even without siblings horning in, the food isn't guaranteed all-you-can-eat. If you polish off your brood ball, there's no kitchen to call for another one. In order to cope, dung beetle larvae have evolved the ability to metamorphose at a range of sizes, so they can still complete their life cycle even if they run out of food early.

And although a nice deep tunnel does protect baby beetles from the damaging stress of temperature variation, not all mothers get a chance to make their tunnels deep enough. The possibility of digging a deep tunnel only exists where adult competition for dung is low, as is the case in the eastern United States. There, says Moczek, "Mom has all the time she needs. She can take half a day to dig a tunnel. In Western Australia, if she takes half a day to make a tunnel, the dung is gone. There are hundreds of individuals competing with her. Mom doesn't have the time to make a deep tunnel, so she makes a shallow tunnel. The price she pays is that larvae are now exposed to temperature fluctuations." Thus, Australian larvae suffer a more stressful environment than American larvae—but evolution is always working. No system in the world remains static, and over time Australian larvae have adapted to tolerate changing temperature better than American larvae, so they may now have an advantage in a warming world.[15]

Moczek's work has shown that organisms don't merely experience their environments, they help create them. Perhaps most critically, parents can pass on a curated environment to their offspring, along with their genetic material. Nature and nurture intertwine in the development of animals from humans to dung beetles. Cow pies, it turns out, are a rich source of both odors and insights into big questions of ecology and evolution.

Like novice parents who learn that the compelling cuteness of a newborn is linked to the relentless reality of dirty diapers, biologists often start off with an interest in nature that leads inexorably to a fecal fate. Emilie Snell-Rood is one such scientist. She began her career with a passion for birds and conservation, then eventually found herself digging insects out of poop. An alumna of Moczek's lab who is now a professor at the University of Minnesota, she says, "When I was a bird person, I always thought of insects as bird food. But as I started to study them, I fell in love. There are as many species of dung beetles in the world as there are birds. I never gave them enough credit—larvae especially." She became entranced watching dung beetle larvae under the microscope. "They're so gross and so fascinating. You can see the dung moving through their body, and they're shaped in a way that they can eat their own poop, which is likely to increase their microbiome, to digest the poop they're feeding on. You can see them digesting poop, they poop, they eat it—it's a cycle of poop."[16]

You might find yourself cringing at the all-too-real image of babies eating poop. Although it's not a great idea to let curious human kids shove cat litter in their mouths, it's worth noting that poop consumption is a common baby behavior throughout the animal kingdom. Koala mothers even produce a special poop called *pap* to feed their young, passing on the good bacteria that their joeys will need in order to digest a diet of eucalyptus leaves.[17] (Between *pap* and *pedestal*, it would seem we need to distance ourselves from edible poops by giving them special names.) Rabbit mothers, too, deliver only two things to their litters during brief daily visits: milk and fecal pellets. The babies, called kits, concentrate on milk for the first week of their lives, then add the feces to their diet in the second, and by three weeks of age they're ready to leave the burrow, armed with the right gut bacteria to help them digest the local roughage.

Symbiotic microbes that pull developmental strings

Microbes are so important to the development of some animals that they can't wait until hatching or birth to get their microbiome. Many of these species actually deposit symbionts inside their egg cells. In 2014, scientists discovered for the first time an animal that could not even consistently complete its first cell divisions without bacteria guiding the cellular apparatus. The animal is a roundworm, the bacteria belong to the genus *Wolbachia*, and if the eggs are treated with antibiotics, about half of them die.[18]

When I learned this, I was amazed and aggravated in equal measures. In the early 2000s, when I was struggling to produce viable squid babies *in vitro*, my colleagues and I tried adding antibiotics to the water. Our hope was to combat the infections that often seemed to kill the embryos before they could hatch. It never occurred to me that we might be killing off beneficial bacteria instead. Eventually, I realized that the infections were likely fungal rather than bacterial, and if bacteria had been allowed to grow, they may well have kept the fungi in check. I'd like to say that none of this was known at the time I was doing my experiments, but research as early as 1989 found that shrimp embryos attract bacteria that produce antifungal compounds. Shrimp denied these bacteria are quickly killed by fungus.[19]

Of course, microbes are so diverse that there's never a single best approach. Many bacteria *can* infect and kill developing babies. It's enough of a problem that some bird mothers put antibacterial compounds into their eggs, while some fish fathers surround their eggs with antibiotic glue.[20] These parenting techniques help make up for the fact that early embryos do not yet have an immune system.

But keeping out the "bad guys" is only one facet of the immune system. It also needs to recognize and negotiate with the "good guys." For many species, finding beneficial symbionts is one of the most crucial tasks of development. When the life cycle of an organism goes through the "single-cell bottleneck," it's sort of like taking a broad-spectrum antibiotic. You've gotten rid of the bacteria that were making you sick—and also the ones that were keeping you healthy. And as we saw with Venom and other symbionts, in many cases this is a muddy distinction. The difference between a parasite, a mutualist, and a commensal is not necessarily in the species but in the environmental context.[21]

Consider the bacterial genus *Wolbachia*, members of which inhabit not only the roundworms we met above, but a huge variety of insect species. The relationship between *Wolbachia* and its hosts is clearly interdependent, but is it "good" or "bad" or a little of both? The most reliable way for *Wolbachia* to get into new hosts is when a female host lays eggs, so male hosts are far less useful to the bacteria. Thus the little symbiont has evolved a variety of tricks to skew the sex distribution of its hosts toward producing more females. In at least two species, one butterfly and one lady-bug, male embryos that inherit *Wolbachia* from their mothers are killed by the symbionts before they can hatch.[22] Pill bug males get off easier; *Wolbachia* simply causes them to develop into females. This is particularly impressive because the sex of pill bugs, like that of humans, is genetically determined.

Pill bug sex chromosomes are, in a way, opposite to our own. A human with two of the same sex chromosomes (XX) develops female traits, whereas a human with two different sex chromosomes (XY) develops male traits. In pill bugs, a pair of the same chromosome (ZZ) produces male traits, while a pair of different chromosomes (ZW) produces female traits. Symbiotic *Wolbachia*, however, can override genetic sex, causing ZZ individuals to develop ovaries and viable eggs exactly like a ZW individual. Over time, this has led to some pill bug populations in which every single individual is ZZ, and the difference between male and female is simply whether or not you're infected with *Wolbachia*. Going one step further, some pill bugs have even incorporated the "feminizing" genes from *Wol-bachia* into their own genome, creating an entirely new genetic sex determination system apart from the chromosomal approach their ancestors took.[23] When biologists discover that something like this is possible, they immediately begin to wonder how many times it might have happened before. In the hundreds of millions of years of animal life on Earth, how often has an intimate developmental relationship with microbes changed the course of a species' evolution? Certainly more often than we know.

Wolbachia's capacity to mess with the development of its hosts appears limitless. Ants, wasps, and bees have a different kind of sex determination, as we saw in the last chapter—unfertilized eggs become male; fertilized eggs become female. But *Wolbachia*, undaunted, has evolved the ability to duplicate the genes of an unfertilized egg. Now the embryo has two copies of every gene, as if it had been fertilized, and so it develops into a female. Is *Wolbachia* unique among microbial symbionts, or is it representative of others that we have yet to characterize?

Environmental dangers and coping with them

We do know that *Wolbachia* is hardly the only environmental factor that fools around with sex determination. Many inanimate pollutants do as well, to the point of threatening the survival of affected species. Atrazine, the world's most widely used weed killer, can cause male frogs to develop ovaries instead of testes.[24] Although it's only deliberately applied to agricultural areas, wind and water runoff distribute atrazine widely throughout the environment. For decades people didn't worry about it because it didn't seem to have negative impacts on adult animals. However, research in the early 2000s provided evidence that atrazine significantly alters the production of the sex hormones testosterone and estrogen during frog development. (These frogs were the model organism *Xenopus*, illustrating that lab studies on model species can be effectively used in environmental research!) Male tadpoles exposed to atrazine not only developed ovaries as they matured but also failed to develop the typical large male larynx for mate-attracting calls. Unlike the few species of parthenogenic lizards, frogs can't reproduce without both males and females. If males are lost, so is the population, and, in the worst case, the entire species.

Climate change is another source of environmental disruption to animals with temperature-dependent sex determination. The idea of incubation temperature determining sex might seem strange, but Aristotle thought that even human sex was determined by temperature. He suggested that people copulate in the summer to generate male heirs. We've since learned that human sex is genetically determined—but if we were like red-eared slider turtles, Aristotle's advice would have produced the opposite effect! Slider eggs incubated warmer than 88°F (31°C) produce female turtles, and colder than 79°F (26°C) produce males. Between these temperatures, embryos can go either way. High temperature apparently activates a gene that produces estrogen, which turns proto-gonads into ovaries. Without that gene activated, the gonads become testes. However, that mechanism isn't universal to animals with temperature-dependent sex. In the American alligator, both high and low temperature extremes hatch females, while intermediate temperatures produce males. Scientists are still puzzling out why.

In many other species, temperature determines not the sex but the speed of development. Parents seek warmer egg-laying habitats to make their babies hatch faster. Sharks and rays, species that as adults have little to do with tectonically active habitats in the deep sea, have been found to deposit their egg cases at cracks in the seafloor called hydrothermal vents. The chemicals spewed from the vents feed specialist bacteria, along with animals like worms and clams that have evolved partnerships with those bacteria. Shark embryos aren't known to have any such relationships, but the higher water temperature around the vents speeds up their development, so adults select it as a cozy incubation site.[25] (In 2018, hundreds of deep-sea octopuses were observed brooding their eggs together near a warm-water vent, in what was inevitably referred to as an "octopus's garden." The observed females exhibited signs of stress from being too close to the warm water, and the observing scientists concluded that even more brooding females must have been hiding out of view, occupying better spots.[26] We'll talk more about brooding in the next chapter, but it just goes to show that good egg habitat can be in high demand.) Deep-water corals and their relatives, sea pens, also provide nursery habitat. Catsharks attach egg cases to them, and redfish hide embryos inside them.[27] Like anemones, both corals and sea pens use sting-ing cells to defend against predators and capture prey, so the fish embryos must have some kind of protection, perhaps similar to the mucus coating of clownfish living in anemones. (Even the details of clownfish protection remain murky, with new evidence suggesting that microbes may be as important as mucus.[28])

One group of deep-sea fish sneak their eggs past stinging ten-tacles with an ovipositor—the same type of structure that wasp mothers use to insert their eggs into caterpillar hosts. These fish, members of a group known as snailfish, use their tool far more indis-criminately than wasps. In addition to corals and sea pens, they'll inject eggs into sponges, crabs, and clams. Snailfish can't be too picky about what to use as an egg nursery, because reliably encoun-tering the same species time after time is unlikely in the sparse habitat of the deep sea. In 2019, scientists found that snailfish will even inject eggs into a type of deep-sea organism called a xenophy-ophore (see insert, photo 4).[29] These organisms are single-celled life-forms, somewhat similar to amoebas, making them part of a group that's usually classified alongside bacteria as microbes. But xenophyophores grow as large as your fist, and they almost seem to have an inversion of a microbiome. Instead of a multicellular organism like a human hosting numerous single-celled species, a xenophyophore is a single-celled organism hosting multiple multi-cellular species—not only snailfish, but several kinds of worms lay eggs inside these giant cells.

Not every animal is lucky enough to have access to a living nurs-ery. Without a ready-made secure location to drop off their babies, many parents build their own. Underwater mothers who lay their eggs in plain sight on the seafloor may devote up to half of their reproductive resources to homegrown egg protection.[30] Amy Moran, a biologist who has studied developmental biology from tropical to polar latitudes, describes the exceptionally sturdy egg case of an Ant-arctic snail. "The capsules were so hard to open," says Moran. "You can imagine a starfish just giving up, 'I don't have time for this.'"[31] Moran did not give up, however, because she wanted to find out how this snail's babies grew so large. Each capsule produced a single juvenile of about a quarter inch (7 mm), larger than a pencil eraser and truly enormous for a just-hatched marine snail. Some huge babies, like ostriches and kiwis, are produced with an abundance of yolk. Moran found that these huge Antarctic snail babies were produced instead with an abundance of nurse eggs. If she opened the capsule early in development, it held a tiny embryo swimming in a banquet of equally tiny nurse eggs. Over time the nurse eggs disappeared as the embryo grew and grew and grew. "That's a different way of making a big juvenile. Same end product."[32]

Moran marvels at not only the toughness but the complexity of egg cases produced by marine snail moms. "They don't have any hands or anything, and yet they make these incredible sculpted multilayered complex things." If you see a repeated row of small solid structures, affixed in rows or spirals to a rock or piece of seaweed at the beach, odds are good that a snail laid it. (While revising this book, I found a lovely series of bright blue capsules arrayed on a strand of seagrass like beads on a necklace (see insert, photo 5). Several experts agreed that they must be cone snail eggs, though none of us have been able to figure out the blue coloration.) These capsules may be shaped like a purse and attached along one seam, or like a wine glass and attached at the stem. Roughly opposite the point of attachment, they have an exit in the form of a capsule plug. Embryos produce a special chemical when they're ready to hatch, which dissolves the plug, opening the door to the rest of their lives. Says Moran, "It makes a chicken egg look like a pretty simple structure."[33]

Parents put a tremendous amount of energy and effort into their eggs, from packing nutrients and microbes to seeking out or creating the ideal nursery. With all that investment in place, it's a major loss if something goes wrong during embryonic development and

3

BROODING EGGS

Carry Them, Sit on Them, Swallow Them Whole

Who can forget the attitude of mothering?
Toss me a baby and without bothering
To blink I'll catch her, sling him on a hip.

—Rita Dove, "Mother Love"[1]

When humans brood, we "nurse feelings in the mind." But not just any feelings—the word has long held a negative connotation. We don't brood over a successful presentation or a gold medal. We brood over hurts and slights and lost opportunities. This kind of brooding is a metaphorical reference to the brooding behavior of hens, when they sit on eggs to keep them warm and safe. Eggs are wonderful things, full of life and nutrition and potential microbial partners! So why don't we use the word *brood* for happy feelings?

Probably because of the way brooding hens act toward people. They'll peck and bite to keep our hands away from their babies; they can even go overboard with their protective instinct and steal eggs from other hens to add to their own. It doesn't make for very pleasant company. Stuck in our human perspective, we started using "brood" to mean worry and mope and sulk instead of releasing our self-absorption, getting into the hen's point of view, and turning "brood" into a synonym for care and tend and foster.

Biologically speaking, hens are simply one of an enormous array of brooding species, including everything from bird butts on nests to fish mouths full of eggs. In fact, brooding behaviors are found on every branch of the animal family tree. Some sponges—animals so simple they have no organs at all, not even a stomach—keep their larvae in special brood chambers. Copepods, little shrimplike creatures abundant in both fresh and salt water, attach their eggs to their tails. Meanwhile, mother pram shrimp hollow out a large gelatinous animal called a salp and use it as a nest for their young.[2] Even leeches can carry their babies in a variety of ways. (From the same word root as pouch-bearing marsupial mammals like the kangaroo, the leech with a pouch is named *Marsupiobdella*.)

Brooding eggs is a lot of work. Beyond the initial investment of building a nest (or gruesomely carving one out of another living animal), there's the constant task of keeping eggs warm, or wet, or well-oxygenated. Different species have different needs, and the world contains an indispensable diversity of brooding habitats. Lizard mothers hunt for a location that's warm but not too warm, while bird mothers use their own body heat. Spraying characid fish lay their eggs out of water to keep them safe from aquatic predators, then become dedicated egg splashers to keep them from drying out.[3] Glass frogs, meanwhile, maintain the moisture content of their eggs by urinating on them.[4] Octopuses attach their eggs to rocks and aerate them with jets of water; crabs hold on to their eggs and flap their abdomens to give them oxygen.

Few natural phenomena attract our attention and admiration as much as parents caring for their babies. We often reach out to help, wanting to build birdhouses for our feathery friends to nest in, or to plant milkweed for monarch butterflies to lay eggs on. But our efforts can backfire if we don't understand the full range of ecological interactions throughout an animal's life cycle. Some birdhouses can increase the risk of predation on eggs and nestlings, for example, but cultivating a pesticide-free yard full of native plants encourages birds to nest naturally.

Sometimes, too, they surprise us.

Brooding on land—safety, warmth, and food

Near my house on a pedestrian freeway overpass, a hummingbird nest fills a single link of a hanging chain. To us humans, it looks like an incredibly inhospitable home. The surroundings are pure concrete, no trees or bushes or grass. It's right at the intersection of three highways, and the traffic noise is usually so loud you have to shout to be heard. On the overpass itself, bicycles frequently whiz by, and at least one family with young children (mine) is always trying to peek inside the nest.

But to the hummingbird, it must be an excellent location: difficult for predators to reach, a short flight from a park well stocked with flowers, and practically invisible. Most people walking by don't notice it, as the chain hangs well overhead. To be honest, I'm not sure I would ever have seen it if my tall husband hadn't pointed it out, eight years ago. Since then, we've found a fresh nest in the same location almost every year, and although it's too high for us to look inside, he can reach up with his phone to take a picture.

One wrenching year, the photo revealed a dead chick decomposing, but most often there's an egg or a successfully hatched chick, covered with needlelike spines that will fledge into feathers. The little thing looks like a hedgehog or an echidna, except for the long thin beak poking up from the nest. Hummingbird chicks, like many other bird species, are *altricial*—naked, blind, and not much resembling a bird. Sometimes people describe altricial chicks as "helpless," but I find that adjective questionable. It's true that they require far more parental care than *precocial* chicks like ducklings, which are fluffy and mobile. But altricial chicks are incredibly well-adapted to acquire that care, with specific calls to stimulate feeding and colorful insides of their mouths to provide a target (see insert, photo 6).

However, we've never heard the baby hummingbird make a sound. At first I thought its peeps were lost in traffic noise, but then I did some reading and learned that hummingbird chicks keep silent to avoid the attention of predators. Hummingbird parents brood the eggs until they hatch, and continue brooding the chicks until they're feathery enough to stay warm. After that, the parents switch tactics and stay *away* from the nest as much as possible, to avoid drawing any predator's attention.[5] Rabbit mothers take a similar approach, hiding their kits underground and visiting only once a day for the briefest possible feeding. These are both animals with virtually no ability to fight off predators, so there's no advantage for them to stay nearby and try to protect their babies. Their best bet for reproduction is to make sure no one notices the nest. They've evolved short feeding times and quiet babies.

By contrast, for many years my family has been serenaded with the "feed-me" choruses of baby phoebes at our house. At first we didn't know what kind of birds they were, but a close look at the parent's incredibly round body, all black with a white belly, and the little feathery crest like an understated mohawk, allowed us to identify it in one of our bird books. The name *phoebe* supposedly describes the sound of their call, but I can't say I recognize it (maybe because I didn't start life as a bird person). The phoebes first began nesting under our front eaves, a location readily visible from the sidewalk and made even more visible by streaks of bird poop on the wall under the nest. Then, last year, before they arrived at their usual nesting site, a bird-loving neighbor stopped by to chat. (And when I say bird-loving, I mean that this neighbor throws mealworms into the air from her front porch every day, and boy do the birds show up for it. Mealworms, by the way, are baby beetles.) She told me that the phoebes at our house, along with all the other phoebes in the neighborhood, are descended from a pair in her yard. She pointed out the crows hanging around nearby, and said they'd be a danger to the phoebe nestlings. She suggested putting a fake crow in our yard to scare them off.

I'm not one to take sides among wildlife, beyond keeping our cat indoors (he's not wildlife, anyway). I decided to wait and see what the phoebes would do. At first it looked like they might not build a nest at all, but then my daughter noticed a phoebe flying up under the eaves behind the house. Soon enough we were sitting in the backyard, watching a chirpy little chap fly back and forth with mud and dry grass, as a nest appeared bit by bit.

I learned from the Cornell Bird Lab that "The male Black Phoebe gives the female a tour of potential nest sites, hovering in front of each likely spot for 5 to 10 seconds. But it's the female who makes the final decision and does all the nest construction."[6] From this I surmise that our male showed the female the front of the house, and she replied "Nope, crows," so they moved to the back.

A cell phone photo of the inside of the completed nest revealed three beautiful blue eggs. The parents did not appreciate our surveillance. They would swoop and shout if we got too close, but we found that we could sit in the hammock in our yard, gently swinging, and they would return to their perches on trees or wires, keeping an eye on the nest while scanning the area for tasty bugs. They particularly enjoyed the large compost bin that sits against our back fence and hosts a constant buffet of flying insects. One day, my five-year-old and I decided to name the phoebes Borb, because they were so round. (We had to give them both the same name, since it's impossible to tell male from female.)

As is typical for birds with altricial young, it seemed like the eggs had barely been laid before they were hatching. Anton lifted his camera again and got a blurry photo that showed at least one chick. Borb shouted him back indoors, to which I offered an encouraging, "Good job, Borb! You chased him away!"

The hatchlings grew astonishingly fast; at least, I was astonished by it. Altricial chicks require a huge parental investment, but the payoff is a speedy development to independence. We heard them cheeping, saw the parents provisioning—and the next thing I heard from the kids was, "They're flying! The little ones are flying!" My son saw them fly off one day while I was working, and that was it. No more nestlings. "I wasn't ready," I complained, to the amused sympathy of my children, who are so altricial they won't leave the nest until they're teenagers.

I like to think we can take some small credit for the phoebes' decision to nest with us, year after year. Although we didn't chase away the crows, we keep our yard stocked with abundant nesting material and delicious bird food by avoiding pesticides, planting native flora, and cultivating a fly-filled compost bin. Although I've told them that creating a bird-friendly ecosystem is the best thing we can do, the kids can't help wanting to put up birdhouses. My daughter once purchased and painted one, which now hangs on our maple tree as a lovely decoration that has, thankfully to my mind, remained empty of birds. Birdhouses and nest boxes, no matter how well intentioned, can become "ecological traps" that attract predators and encourage birds to nest in unsafe places or to lay more eggs than they can feed.[7]

Ecological traps can arise for other nesting animals, too, like tropical lizards. These cold-blooded animals can't keep their eggs warm by brooding them, so they seek out warm nesting sites. On Taiwan's Orchid Island, long-tailed skinks typically laid their eggs beneath rocks—until they came upon the availability of concrete retaining walls. In the early 2000s, female skinks that laid their eggs in concrete had greater hatching success than those who laid eggs in the wild, thanks to the greater warmth of the walls compared to the rocks. However, within ten years climate change had driven temperatures high enough that the hot concrete began damaging embryos and reducing hatching success. The rapid pattern reversal illustrates the complexity and variability of human influences on animal development.[8]

Brooding in water—it's all about the oxygen

We animals need oxygen. Or rather, our mitochondria need it, those age-old symbionts-turned-cellular-machinery, and we need our mitochondria. They use oxygen to make energy for everything else we do with our bodies. Thanks to photosynthetic plants and algae, Earth's air is about 21 percent oxygen—plenty to support a profusion of air-breathing animal life. So it's a shock to remember that animals first evolved in the ocean, which even today contains less than 1 percent oxygen.

Now, you might be inclined to call this percentage ridiculous, and point out that every molecule of water is represented chemically as H_2O. Obviously, the ocean is full of oxygen! But it's full of oxygen atoms bound to hydrogen atoms, and this is not the kind of oxygen that living organisms breathe. We need oxygen gas, O_2, whether it's mixed with other gases in the air or dissolved in liquid water. The most thoroughly oxygenated seawater can only hold about 1 percent dissolved oxygen. For a fish to get the same amount of oxygen as a squirrel, it has to breathe far more water than the squirrel does air. Plus, water is more viscous than air, meaning that it takes more work to pump water over gills than to fill lungs with air.

Climate change is exacerbating the situation. Warmer water holds less dissolved gas (one reason that people who enjoy fizzy drinks often prefer them cold—they lose fizziness as they warm up), and the ocean has grown demonstrably warmer in recent decades. In addition, pollution from agricultural runoff has been driving the growth of oxygen-free "dead zones" in areas near the shore. Adult animals can often escape by moving into more oxygenated water, but developing embryos are at the mercy of their environment.

Parents that leave their eggs to fend for themselves, like those we met in the previous chapter, typically produce egg masses that take oxygenation into account. Consider the enormous Humboldt squid egg mass we found in the Gulf of California. In trying to figure out how a mother squid could lay an egg mass so much bigger than she is, we hypothesized that she might create a concentrated mucus that inflates upon contact with water. One advantage to such a theoretical expanding sphere would be to spread out the eggs, making it easier for oxygen to reach them all.

Salamander mothers, who lay water-bound eggs like frogs, don't always rely on diffusion of oxygen from the surroundings. The spotted salamander, *Ambystoma maculatum*, packs an extra source of oxygen into the egg mass itself: symbiotic algae (see insert, photo 3). The specific alga (singular of algae) was named for its starring role, *Oophilia amblystomatis*, "lover of ambystoma eggs." The salamander-alga relationship has been known for over a hundred years, but in 2010, scientists were stunned to discover that the alga actually grows *inside the cells* of the salamander embryos. Although symbiotic microbes have been found inside invertebrate egg cells, like the *Wolbachia* we saw in insects, this was the first time such intimacy had been documented in a vertebrate—an animal with a backbone.[9] Unfortunately, further exploration of this fascinating symbiosis proceeds alongside studies of its vulnerability, as scientists have also discovered that the alga is susceptible to the abundant herbicide atrazine. If the alga is killed, many eggs suffocate and fail to hatch.[10] Thus, a great diversity of developing animal babies suffer the influences of commercial agriculture, from fertilizers to herbicides to pesticides. (We'll hear more about pesticide disruption in the next chapter.)

Given the crucial importance of oxygenation to animal development, especially in the water, many parents have taken on the task themselves. Octopuses, which one researcher I spoke to called "the poster child for good mommies,"[11] lay their eggs in strings or clusters and continuously aerate them with gentle puffs of water. In most species, octopus mothers are so devoted to this task that they stop eating, and die about the same time their eggs hatch. Octopus brooding usually takes a few weeks or months, but in 2014, the marine biology world reeled at the discovery that an octopus mother in the deep sea had brooded her eggs continuously for four years.[12] A few different deep-sea octopus species brood their eggs inside their body, near the gills that they're already pumping fresh water past. Brooding oysters, meanwhile, keep their eggs right *on* their gills, the site of maximum oxygen exchange. Nevertheless, these brooded oyster embryos suffer more oxygen stress than those who are released to develop freely in the water, suggesting that, in the case of oysters, brooding arose as an adaptation to protect the young rather than to help them breathe. Interestingly, when they hatch, these brooded oysters are more resilient to environmental low oxygen.[13] Are they the lucky ones, preadapted to a low-oxygen world? Or will reductions in available oxygen make their already stressful brooding environments unlivable? We don't yet know.

In most cases, mothers who brood eggs inside their bodies are small and so are their egg masses. Holding your children too close runs the risk of smothering them, so you can only do it if you have a relatively small batch of eggs through which oxygen can easily diffuse.[14] But Chilean scientist Miriam Fernández studied a remarkable exception, the brooding crab *Pseudograpsus setosus*. Mother crabs grow up to 5.5 inches (14 cm) and produce egg masses far too large to obtain enough oxygen by waiting for it to passively spread into the mass from the surrounding water. Crabs brood their eggs in a mass under their abdomens, and while small crabs simply hold their eggs in place, Fernández found that large crabs actively flap their abdomens, ventilating their babies. Was this behavior enough to keep the eggs healthy?

To make a clear connection between the mother's actions and the oxygenation of her offspring, Fernández had to measure oxygen inside the egg mass while it was still attached to the mother's body. She was working in Germany at the time, in a laboratory that had plenty of oxygen-measuring instruments—all made of glass. The active movement of the crab mothers would break the

A mother crab broods thousands of eggs beneath her abdomen—typically a safe and hidden place, unless a primate paw picks her up for inspection.

instrument within seconds of inserting it into the egg mass. Then Fernández heard about a new oxygen meter being developed in Bremen, not far from her lab. This instrument used optic fibers with a special coating at the tip that would react with oxygen. As brand-new technology, it was expensive, but the German lab had plenty of money. Fernández found that the vigorous movement of the crabs could still break the meter's tip, but unlike the other instruments, its optic fiber could be recoated. This was a delicate task, and it had to be done often. The creators of the instrument wanted very badly for it to be successful. They taught Fernández how to fix it herself, so she could keep using it. When she moved back to Chile, she brought both the instrument and the coating pigment. It has allowed her to continue this research, although she worries about not having the same resources as in Germany, if she ever has to buy a new optic fiber. I'd hope that the company would make it possible, given that, as Fernández says, "The first optic fiber of this company that now provides optic fibers all over the world was used for my crabs."[15] Her most remarkable discovery was that these large crab mothers could monitor the oxygenation of their eggs in real time, and respond to low oxygen with increased ventilation. The crabs don't have an oxygen meter of their own, but they seem able to detect changes in their embryos' biochemistry, which can be used as a proxy for oxygenation.[16]

The egg-brooding behavior of crabs is just as expensive, in its own way, as the research has been. Instead of costing money, it costs energy. Flapping your abdomen isn't free, and working hard to help your embryos breathe increases your own oxygen needs—a troubling prospect, as oxygen becomes less available in the sea. Layered onto the challenge of reduced oxygen is increased temperature, and animals all need more oxygen at higher temperatures. In warmer water, a mother crab ventilates her eggs more, which makes her breathe faster, which means she needs even more oxygen. Fernández thinks that these crabs will have to migrate as zones of reduced oxygen expand along the Pacific coasts of both North and South America.

Mothers aren't the only parents spending their energy on expensive brooding behaviors. Among sea spiders, it is fathers who collect egg masses from multiple females on their legs, then dance around to keep their babies supplied with oxygen.[17] Many other animal dads make a similar commitment, from midwife toads who wrap egg strings around their legs to water bugs who give up flying while eggs are glued to their backs.[18] Males of the sadly extinct Darwin's frog brooded babies in their vocal sacs, giving up their voice to nurture their young.

Species with brooding fathers have to convince females to hand over their precious eggs, which in stickleback fish has led to the evolution of a strange behavior: egg thievery, where fathers steal eggs that do not belong to them. Females are more likely to choose nests that already have eggs, because if other females have trusted this dad to raise their offspring, he's probably trustworthy. This puts aspiring, eggless fathers in a bind, and so stickleback wannabe dads snatch eggs from other dads' nests to make their own look more attractive.[19]

A number of fish are mouth brooders, which is similar to gill brooding in that it maximizes the oxygen exposure of embryos. In these species, the mother lays her eggs, and then either mother or father scoops up the eggs to carry in their mouth until the babies hatch—and sometimes after. Depending on the species, it's difficult or impossible to eat while mouth-brooding, making this a significant parental sacrifice. But it can also turn into a sacrifice for the babies. A mouth-brooding cardinal fish father will eat the eggs of his previous mate if he comes across a new and preferable partner.

Mouth-brooding cichlid babies face another danger: brood parasites. Sometimes a catfish sneaks its eggs into a pile of cichlid eggs right before the parent scoops them up. Catfish embryos develop more quickly than cichlid embryos—an evolutionary adaptation, not a coincidence. This allows the catfish babies to hatch sooner, and before swimming out of their unwilling foster parent's mouth, they add insult to injury by gobbling up their step-sibling cichlid eggs.

Given how expensive brooding can be, the prevalence of brood parasitism is hardly a surprise.

Cuckoos: birds, bees, and more

Unlike Snell-Rood, the bird enthusiast who grew to love dung beetles, I used to love worms and caterpillars so much that I thought of birds as nothing more than annoying predators. But as I've gotten to know the birds around me, like the nesting phoebes, I've come to appreciate them more. I adore their peeping, fluffy babies, and I'm always curious to find out if a cowbird might parasitize their nest, leaving a much larger and louder baby for them to raise. Ever since she learned about brood parasites, my daughter asks of any eggs we see: "Are these laid in someone else's nest?" Given the frequency of this behavior in the wild, her suspicion is entirely warranted.

"[For] all the animals that have brood care, there's a parasite out there somewhere, parasitizing that care," explains Australian biologist Ros Gloag, who has spent the better part of two decades studying brood parasitism.[20] She began looking at the behavior in honeybees, then moved into South American cowbirds and Australian cuckoos.

I'd heard about European cuckoos, "the ones that got famous in cuckoo clocks," as Gloag calls them, and I knew about cowbirds because I live in North America, where they are the cuckoo equivalent. But I learned recently that cowbirds aren't native to my home state of California, having only moved to the West Coast along with European settlers. The local birds lacked any of the defenses we'll learn about shortly, so cowbird parasitism has contributed to the endangered status of two native songbirds.[21] However, I didn't realize until talking to Gloag that brood parasitism has evolved in birds at least *seven different times*, in seven different lineages. "American cowbirds are actually a type of blackbird," she tells me. "African cuckoos are finches and honeyguides. Oh, and there's a parasitic duck."

Excuse me, a parasitic duck? Here I request a conversational detour to discuss this species. I feel validated to learn that it really is as unusual as it sounds. The black-headed duck, *Heteronetta atricapilla*, is the only duck that never builds its own nest, and in fact it's the only brood parasite with precocial, rather than altricial, chicks. Brood parasitism has evolved more often in birds with altricial chicks, because they expend so much energy raising and feeding their hatchlings. But even making a nest and sitting on eggs is a cost, albeit a lesser cost, and black-headed ducks have evolved to duck out of it. They lay their eggs in the nests of other water birds—mostly coots, who aggressively protect their broods from predators. Within hours of hatching, the ducklings waddle off to make their own way.

Many parasitic species display similar strategies to trick their hosts into providing care. Gloag explains it as an arms race. Cuckoos begin laying eggs in host nests, then the hosts evolve the ability to recognize and remove cuckoo eggs, then the cuckoos evolve eggs that look like host eggs. Hosts become more discerning, and the egg mimicry becomes more convincing, to the point of matching size, shape, color, and pattern, in ways that we humans can't even see.

But one Australian cuckoo has evolved its egg color, not to defend against host recognition but to defend against destruction by *other cuckoo chicks*. The trademark behavior of a cuckoo chick (though not a cowbird chick) is to push any other eggs or chicks out of the nest as soon as it hatches. It's quite a sight: this blind and naked wretch of a thing, wriggling around with its head down, using its legs to evict nestmates. That includes both the children of its host parent and the children of any other cuckoos who parasitized the same nest. So, when a female cuckoo is looking for a nest to lay in, if she finds other cuckoo eggs already present, she'll break them or roll them out to make sure hers is the first to hatch. That's bad news for the other cuckoo moms, of course. To hide their eggs from fellow cuckoos, the Australian species has evolved dark brown eggs that blend into the darkness of the nest, rather than white speckled eggs that would match those of the host species.[22]

The hosts of these Australian cuckoos, rather than attempting to reject cuckoo eggs, have evolved strategies to reject cuckoo chicks. This, too, is a unique behavior among parasitized birds. "Sometimes we see the host pick up a chick that looks almost identical to their own chicks and toss it out of the nest alive," says Gloag. "Which is not a thing mother birds do, generally." The arms race goes on, and now the cuckoo has evolved chicks that hatch looking just like baby hosts, despite a complete lack of resemblance between the adult birds. The cuckoo mom is three times bigger than the host mom—but the cuckoo chicks look so much like host chicks it's hard even for researchers to tell them apart.[23]

Mimetic chicks have also evolved among the cuckoo finches in Africa, but in this case, the mimicry helps them beg more effectively for food. The inside of a baby bird's mouth, its *gape*, is often brightly colored to get the parents' attention. Finch chicks have the most diverse and striking gapes of all, with different patterns in each species, and cuckoo chicks mimic the precise pattern of their hosts.

The South American screaming cowbird (which may have the best name of any brood parasite) has evolved yet another type of mimicry: fledgling mimicry. The fledgling stage takes place once young birds have learned to fly and left the nest, but while they are still dependent on their parents for food. Screaming cowbird chicks don't need to look like host chicks, because the parents automatically feed whoever's in the nest. However, once they're out of the nest they need to be recognized to be fed, so they grow plumage to look just like host fledglings. They even make a call that mimics those of the host fledglings.

Another South American cowbird, the shiny cowbird, has hacked the host parents' provisioning system to create a begging call that's even more compelling than that of the species' own young. These cowbirds use dozens to hundreds of different hosts, but their call sounds nothing like the chicks of any of these species. Instead, it somehow manages to send a message that can be received and interpreted by all kinds of birds. Gloag once played a recording of this call to a bird in the United Kingdom, who had never seen or interacted with a cowbird in its life or in the evolutionary history of its species, and it responded as if hearing its own chick.[24]

Cowbird chicks have another auditory advantage when it comes to parental provisioning. Unlike cuckoos, they've allowed their nestmates to survive. At first this might sound like unwelcome

competition, until you understand that bird parents work harder
to provision a nest containing more chicks. The more begging a
parent hears, the more provisioning it does. And because cowbird
chicks are bigger than their host siblings, they can throw their
weight around when the food arrives. Says Gloag, "It's like: get all
your siblings to scream for an ice cream, so that your parents give
in and go buy five ice creams, but then you eat four of the five ice
creams."

As humans, watching all of this go down, it's hard not to see the
host parents as dupes. We perceive them feeding a parasite that's
obviously not their child, and we laugh sadly. However, none of us
animals are actually experts at recognizing our own children right
off the bat. We learn, and learn quickly, but there's generally an
element of circumstance involved. Emperor penguin dads spend
months incubating their eggs, but when the egg hatches, they
have to learn their baby's call in order to recognize it when they
come back from fishing. If you swapped out a different baby at the
moment of hatching, the dad would learn that baby's call instead.

Gloag addresses "that classic image of a tiny little red war-
bler feeding a monstrous chick that looks nothing like them" by
reminding me that the behavior is completely normal. The default
for any bird parent is to feed the chicks that are in its nest. "I think
we assume that recognizing your own offspring should be something that nature does, but it really doesn't. Humans can't do it
either, really."

At first, I was shocked to hear this. Then I gave it some more
thought. I recognize my children because I've been caring for them
since birth. At birth, at the first encounter with each baby outside
of the womb, how did I know it was my baby? Not through smell or
taste or appearance. I knew because it had come out of my vagina.
If someone else's baby had done that, you'd better believe I would
have considered it mine. (Although, as one of my friends pointed
out, I might have been suspicious if it looked like a truly different

species.) That's similar to the experience of host bird parents when another bird's egg hatches in their own nest.

Over time, individual birds do learn to be cannier. Research shows that older hosts are more likely to identify cuckoo eggs correctly, or to give up on a parasitized nest and try again elsewhere. What's more, they may become less likely to be parasitized in the first place, developing skills to hide and defend their nests. "It's actually pretty tough going for the parasites," says Gloag. "Cuckoos tend to cycle through hosts. We get to observe a tiny snapshot of evolutionary time, what is happening right now, but if we could zoom out and look at hundreds of thousands of years of time, we'd see a bird like the common cuckoo, it will parasitize reed warblers for some period of time, until the reed warblers develop defenses that the cuckoo cannot overcome. But no drama for the cuckoo, because in the meantime it's also parasitizing buntings, whose defenses are not quite as good, and then they become the main host in that region. Then the bunting defenses get really good, and they switch back to the reed warblers, who've lost some of their defenses. Common cuckoos we think have been doing their thing for maybe twenty, thirty million years. They're good at it."[25]

How did it all start? Probably as soon as parental care itself evolved. The selection pressure for parasitism arises automatically from a capricious environment—suppose a storm destroys a bird's nest right before it's ready to lay. Why not lay in a neighbor's nest?[26] This kind of *opportunistic* brood parasitism can happen between members of the same species or between different species, and it's extremely common. Remember Moczek's dung beetle mothers, and all the effort they put into rolling up brood balls and digging tunnels? He's found that female dung beetles will regularly parasitize each other's tunnels and brood balls, both within and outside their species. They don't just add their own egg, they dig into the brood ball to kill the host's egg, then lay their own. The tougher the environmental conditions, the more likely they are to

parasitize each other. In the laboratory, Moczek has shown that the
rate of brood parasitism rises from 10 percent up to 50 percent as
he makes conditions more stressful.[27] Ground bees, who dig tunnels similar to dung beetles and provision them with brood balls
made of pollen instead of poop, also parasitize each other. Futher-
more, thousands of species of cuckoo bees specialize in parasitizing
both ground bees and the more familiar social bees. Did the first
carrion beetles back in the Cretaceous parasitize each other's carefully prepared dead dinosaurs? Probably.

"You can find everything in insects," Gloag assures me. Numerous butterfly species, for example, parasitize the brood care of ants
by producing caterpillars that mimic ant larvae. They don't even
have to lay their eggs in an ant nest. The caterpillars smell so much
like baby ants that they attract workers who assiduously collect the
"lost" larvae and bring them back underground. Inside the ant colony, they're protected and fed (see insert, photo 7). Some species
will even devour their adopted ant siblings, consuming so many
larvae that the colony collapses.

But of all brood parasites, my favorites are the flies. Remember those parasitoid wasps that lay their eggs inside other insects?
Instead of targeting caterpillars that continue to crawl around and
munch on leaves while the wasp babies grow inside them, some
wasps paralyze a large prey item, bury it underground, and lay a
single egg upon it. Kleptoparasitic flies have evolved to take advantage of this situation. A pregnant fly mother finds a wasp mother
who's carrying prey and follows her, flying in her blind spot. Then,
after the wasp hides the prey and lays her egg, the fly sneaks in to
deposit her own offspring. Because she gives live birth rather than
laying an egg, her larva can immediately eat both the wasp egg and
the prey collected by the wasp mother.

One reason I love this arrangement so much is that parasitoid
wasps horrify me, and so their setbacks delight me. Another is
that it involves fly pregnancy, which I find fascinating. One of the

strangest facts about flies is that they've evolved live birth more often than any other group of insects. If you didn't know that any insects at all had live birth, don't worry—most people don't. But it's true! Some flies not only can get pregnant but also produce a kind of milk to feed their young. For that matter, so do cockroaches.

Pregnancy could be considered the ultimate protection against brood parasitism. No animal—other than humans—has yet figured out how to implant its own embryo into another animal's womb. Pregnancy may have evolved as a protection against parasitism, both brood parasitism and infectious disease, but it comes with its own hazards, as we'll see in the next chapter.

4

PREGNANCY
Not Just a Mammal Thing

His own parents,
He that had father'd him, and she had conceiv'd him in her
    womb, and birth'd him,
They gave this child more of themselves than that;
They gave him afterward every day—they became part of him.

               —Walt Whitman, "Leaves of Grass"[1]

In 2012, four years after my summer of intensive invertebrate baby-making, I finally tackled the project of growing a human embryo. I got pregnant at the beginning of the year and spent my free time for the rest of it reading avidly about human development. I imagined my embryo's cells dividing and specializing into different kinds of tissue. I took folic acid to help the nervous system develop correctly. I often wished that I could see what was happening, the way I had watched my squid babies develop under a microscope. As I grew rounder and more ungainly, I also wished that I could have laid an egg early in the process, which my spouse and I could now take turns incubating. When birth approached, with its attendant pains and possible complications, it also occurred to me to wish that I were a marsupial, birthing a minuscule infant who would then crawl into my pouch all on its own. Even if I were a chimp, I'd have a far smaller baby head to push through the same size birth canal. As it was, my babies' skulls were squeezed in transit for so long that they were notably cone-shaped on emergence.

(Fortunately, humans have evolved unfused skull bones to accommodate both the trials of birth and the rapid brain growth that occurs afterward.)

Since my obstetric ordeals, however, I've read about many pregnant animals with far more traumatic birth experiences than humans. Hyenas give birth through a long narrow channel inside their phallus (yes, females have one) which inevitably tears. A tsetse fly has to push out a baby nearly the size of her own body. As for physogastric mites . . . well, let's save those for later.

When it comes to feeding babies with our own bodies, we humans don't skimp. During pregnancy, we invite embryonic tissues to literally invade our own, forming a placenta and siphoning nutrients from our blood. After birth, we can spend years making milk, tailored to the baby's needs at every stage of development.

We are extremely dedicated— but we are not *exceptionally* dedicated. Placental pregnancy is found in all placental mammals, from cats and dogs to capybaras and dugongs, but it has also evolved in certain species of sharks, lizards, frogs, worms, snails, and insects.[2,3] Milk is all over the place, too. Pigeons, penguins, and flamingos all make crop milk to feed their chicks, jumping spider moms secrete milk from their abdomens,[4] and one cockroach feeds its young with a high-protein milk that, strangely enough,

Parents across the animal kingdom produce nutrition we might as well call "milk," whether it's secreted from a monkey's nipples, coughed up from a bird's crop, or oozed from the walls of a shark's uterus.

offers promising medical applications for people. Then there are animals like earwigs who are even more serious about feeding their children, allowing them to eat directly of a mother's or father's flesh, even unto the parent's death. These species showcase the mutual adaptation of parent and child to each other's needs, a team effort to produce the next generation.

For many animals, parents constitute an individual's first habitat and first food—an entire ecological niche, lasting days to years. The womb may seem like an isolation chamber, protecting its embryonic contents from the outside world, but in fact this microcosm can reflect and even magnify influences of the global ecosystem. External nutrients and chemicals percolate through blood, placenta, and uterine walls to influence development along with the embryo's internal genes and proteins, from the very beginning of pregnancy.

The parent as environment

We've already met several internally brooding fish, and aquarium hobbyists are familiar with live-bearers. Most of these are *lecithotrophic*, meaning that the embryos live off yolk while growing inside their mother. In the well-known case of the seahorse, a father receives eggs into a pouch on his body, which he then seals (see insert, photo 8). The embryos carry yolk from their mother to sustain them, although their father supplements their needs with extra nutrients, hormones, and oxygen. Lecithotrophic pregnancy is found in fish ranging from tiny guppies to the largest fish in the world, whale sharks. (Not a recommended species for aquarium hobbyists, or even aquarium professionals.) Some shark embryos begin by consuming their yolk, then turn on their fellows. Inside the womb, mako shark embryos eat unfertilized nurse eggs, and sand tiger sharks eat their siblings.[5] This behavior is a kind of adelphophagy, like the sibling cannibalism we saw previously in worms and snails.

Do all these cases count as pregnancy, or are they merely internal brooding? The definitions aren't exact. Words like *brooding* and *pregnancy* have been used by humans for far longer than we've known about the full range of possible baby-care strategies in the animal kingdom. With every new discovery, we have to figure out whether it can be slotted into an existing category or demands the creation of a new category. For example, the eggs of certain live-bearing fish hatch while still in their mother's oviducts—structures like our human fallopian tubes, which usually serve as conveyor belts delivering eggs to their final destination. These fish embryos proceed to munch on a thickened lining inside the oviducts for their first meals. What terminology describes this type of parenting?

A useful term that encompasses a range of behavior is *matrotrophy*, or mother-eating. This rather brutal word describes an embryo getting nourishment directly from its parent—typically the mother, but cases of patrotrophy exist as well. Matrotrophy includes taking up a nutritious substance secreted by the parent, often referred to as milk, whether it comes from the teats after birth (breast milk) or from the womb itself during pregnancy (uterine milk). Animal babies that absorb nutrients through a direct connection (placenta) or eat parts of the parent's body for breakfast are also matrotrophic.

Lecithotrophy (yolk-eating) and matrotrophy can be combined in the same species, complicating our attempts to categorize. Sharks and their close cousins, skates and rays, employ just about every technique in the book. Great white sharks sustain their babies with a mixture of nurse eggs and uterine milk, a substance full of fats and proteins that the mother secretes into her womb.[6] Blue sharks, bull sharks, lemon sharks, and several other species take the placental route.[7]

A placenta mediates the exchange of gases, nutrients, and wastes from the mother's body to the offspring, and it constitutes part of the *conceptus*, or material derived from the product of conception (a fertilized egg). Thus, the placenta is grown from embryonic tissue, not maternal tissue. Placentas, in fact, are extremely similar to yolk sacs. Both connect to the bellies of embryos, and both leave behind an "umbilical scar" when they're disconnected at birth. Yup, birds and reptiles have belly buttons, too! However, unlike mammals, they typically heal these scars so well that they disappear in early life. One remarkable exception was uncovered in 2022: the 130-million-year-old fossil of a dinosaur that had retained its belly button to adulthood.[8] This similarity in form and function can help us understand placental shark embryos, which begin by consuming yolk from a yolk sac. They use it up quite early in their development, and then the empty yolk sac attaches to the wall of the uterus to become a placenta. The short stalk connecting yolk to embryo becomes an umbilical cord. Et voilà, a placental shark! Several species of lizards, snakes, and amphibians also have placental embryos, and it's been estimated that matrotrophy in general has evolved at least thirty-three separate times in vertebrates.[9]

That's startling enough. Let us now consider matrotrophy within the *invertebrates*, animals without backbones. Scorpions as a group are entirely live-bearing, with some lecithotrophic species and some matrotrophic species. Within the matrotrophic scorpions, some drink uterine milk and others possess a kind of placenta, although their anatomy and development are different enough from vertebrates that it's hard to draw a precise analogy. (The analogy grows stronger after birth, when a scorpion mother catches her babies and moves them to her back. With her brood thus positioned, she looks just like an opossum mother dripping with babies, or indeed a human mother giving piggyback rides.) We've already met some live-bearing flies—the ones who lay their larvae on wasp prey. Although they parasitize wasp parental care after birth, they invest a huge amount of energy before birth. These mother flies provide enough uterine milk to grow an embryo that fills their entire abdomen. Fortunately they're able to birth these enormous babies without significant damage to themselves, and they can go on to repeat the performance.[10]

Physogastric mites are not so lucky. To be *physogastric* is to have a huge balloon of an abdomen for the purposes of reproduction. For example, you may have seen termite queens, which look like the normal front half of a termite attached to a giant white blob. The termite queen's abdomen is full of eggs, which she will lay over many years, to be cared for by workers as they develop to hatching. A mite mama, by contrast, has a swollen abdomen full of eggs which hatch while still inside her. So far, that's not too different from some lecith-

She may look like an insect taped onto a balloon, but a physogastric mite's swollen abdomen is as much a part of her body as legs or eyes.

otrophic animals. But then, *still inside their mother*, these offspring develop to full maturity, mate incestuously, and burst free, thereby ending their mother's life. The few sons in the female-biased group of siblings soon perish, while the daughters move on to find food and grow their own already fertilized eggs.[11]

The disturbing (to us) reproductive habits of these mites can have significant economic impact. One physogastric mite species is a serious pest of mushroom crops, and insight into its reproductive cycle is necessary to control it. Another species preys on the lesser mealworm, which is itself a pest of human crops. This second species, if encouraged to reproduce even more, could be an effective biological control of the mealworms. What I find most fascinating is that the female mites eat mealworms while they are still eggs. A pregnant mite finds a mealworm egg and sucks it dry, using the mealworm embryo and yolk as nutrients to feed her own offspring. Thus, a single mealworm egg sustains an entire family of mites.

It's amazing how many different ways adult animals have evolved to be, essentially, baby-growing environments. The needs of animal babies have driven incredible adaptations in their parents, from wombs to edible skin to physiogastry.

From one perspective, the mother mite sacrifices herself for her children. From another, her death occurs entirely in her own interests, passing on her genes to the next generation. After all, this is evolutionary biology's definition of fitness: reproductive success. But are parent and offspring interests always well aligned? How can we tell, and what if they aren't?

Are babies parasites?

The most controversial thing I've ever put on the internet was a blog post titled "Why Babies Are Not Parasites." I was happily pregnant at the time, after two years of trying unsuccessfully to become so, and I was feeling kindly toward my burgeoning offspring. Being a biologist, moreover a biologist whose first research adviser was a parasitologist, I was familiar with certain similarities between fetus-mother interactions and parasite-host interactions—the suppression of the immune response, the battle over nutrition. But I couldn't get past the basic evolutionary definition of a parasite as a symbiont that reduces its host's fitness. As I argued in the post, a baby, by definition, increases its parent's fitness.

If you've ever tried to make what you think is a straightforward argument on the internet, you're well aware of the fiery discourse it can ignite. Luckily, my blog's readership was minimal, so I didn't have the honor of experiencing anything like a proper pileup. But even years after posting, my parasite piece would still get the occasional comment from a stranger, and it was always a fierce rebuttal. Babies *are* parasites, my erstwhile commenters would insist. It's in the *science*.

Ten years into my parenting journey, I think I'd write a different post. For one thing, research into mammalian fetal development, immune systems, and the interplay between them has come a long way. That's a lot of grist for thought and discussion. As we've mentioned already, when it comes to symbiosis, the line between a positive and a negative interaction is blurry and shifts depending on environmental conditions. Bacteria can be beneficial in one part of our body and detrimental in another. Eggs raised by an unrelated parent can hurt the host parent's fitness, as in the case of a cuckoo chick pushing host chicks out of the nest, or they can help the host parent's fitness, as in the case of a stickleback father gaining more eggs to fertilize because he showed off the ones he stole.

Like many hosts, a pregnant parent does mount an inflammatory response to the fetus, and like many parasites, the fetus has evolved an array of tricks to prevent rejection. And yet, as eminent developmental biologist Scott Gilbert points out, these adaptations have been coevolved by both parent and child. "It's like the uterus says, *Yeah, I recognize that parasite metaphor. I don't like it. I'm going to negotiate with that parasite metaphor.* I don't think there's antagonism. *You've got your agenda, I've got my agenda, and the bigger agenda is to keep a life going for nine months.*"[12]

Gilbert's perspective is informed by a lifelong passion not only for biology but for philosophy and history of science. Though early in his career he was determined to study genetics, along the way he became captivated by the wonders of embryos. To watch an embryo develop is to witness what is perhaps the world's best rebuttal to the reductive idea that genes alone define an organism. As the embryonic cells divide, each cell inherits a copy of exactly the same genes, and yet somehow environmental signals (even if that environment is mostly its fellow cells) cause certain genes to be turned on or off in certain cells, producing a profusion of differences among blood, bone, brain. In graduate school, Gilbert spent a summer at the Oregon Institute of Marine Biology, another West Coast powerhouse of marine embryology, located midway between Monterey and Friday Harbor. "Everyone there knew that if they found an embryo in their seawater to give it to me. I wrote a paper on the aesthetics of embryology."[13]

I listen, entranced, while he waxes philosophical. As Gilbert describes it, the genetics aesthetic is abstract. Organisms are represented as models, and mathematical predictions of inheritance take center stage. By contrast, the embryology aesthetic is Romantic, in the historical sense. The German Romantic tradition provided two words for development: *Bildung* and *Entwicklung*. Bildung is coming into being, and we see it in the *bildungsroman*, or coming-of-age story. Entwicklung is also used for photographic development, and it represents a potential made actual. Both words focus on the real-world process of building an organism from scratch.

Gilbert became such an established authority on development that he wrote the literal textbook, *Developmental Biology*, first published in 1985 and regularly updated for each new generation of students. By the early 2000s, when I was reading it as an undergraduate, Gilbert was grappling with an increasingly troubling lacuna in the book and in the field as a whole: environmental effects. He credits a Dutch historian of science who pointed out that all mentions of the environment in his textbook were "marginalized" into "sidelights and speculations." "I said, I don't know if there's a coherent framework that would organize them," recalls Gilbert. "She told me, Find one."[14]

This conversation, and a subsequent deep dive into both historical and contemporary research, birthed Gilbert's 2015 textbook *Ecological Developmental Biology*. The text emphasizes that the vast majority of developmental biology studies have been conducted on a very small number of organisms: fruit flies, frogs, chicks, rats. They grow easily in the laboratory, which makes them great for research but hardly representative of animal life as a whole. In their ease of adjustment to laboratory life, model organisms aren't actually models—they're oddballs. However, in recent years researchers around the world have been returning to the early embryological tradition of studying development in the wild, or at least in more realistic laboratory conditions, as well as investigating a greater range of species. We're now gaining developmental insights from organisms as diverse as beetles, butterflies, worms, and octopuses.

A more nuanced understanding of environmentally influenced development comes back to bear on the subject of human pregnancy, with recent work on the interaction of the developing human with its uterine environment. "How the uterus is able to keep the embryo for nine months is absolutely mind-boggling," says Gilbert. The initial arrival of the embryo into the uterus, after it's traveled there from the site of fertilization in the fallopian tube, instigates the same kind of immune response any foreign object would. That sounds pretty parasitic—but then the uterus takes action. Normally, white blood cells would be summoned to attack the invaders, but uterine cells block this part of the reaction.[15] Gilbert describes the results with delight and amazement. "The inflammatory response stays local, and instead of being dangerous, it softens the uterine tissue. It's turning swords into plowshares."

The embryo and the uterus together are engaged in a kind of niche construction, at once entirely different from and analogous to the niche construction of a baby dung beetle and a brood ball. The brood ball is an environment prepared by the mother and seeded with microbes and worms. The grub engages with that environment, actively altering it by mixing in its own poop even as it eats and grows. Like the beetle grub that seems so isolated and protected from the environment outside its ball, the human embryo seems like it ought to be well-insulated from any conditions external to the uterus.

And yet.

Danger crossing over

During pregnancy, embryos are vulnerable in ways that adult animals are not, both to intrinsic and extrinsic factors. If environmental toxins are going to mess you up, the risks are much higher if you start accumulating them while still an embryo than if you don't encounter them until you're an adult. And if your genes are going to kill you directly, by building part of your body wrong or not building it at all, that's most likely to happen at the very beginning of your development—although how often it happens is a hard question to tackle.

Eggs, whether laid in a nest or brooded, whether surrounded by a hard shell or a soft membrane, can be collected and examined with comparative ease. Still, it was only recently that Hemmings developed a technique to uncover the hidden percentage of early bird embryos that don't survive. Early embryo mortality in pregnant animals is probably also quite high, but even more difficult to study. In humans, estimates of the percentage of conceptions that are spontaneously aborted range from 31 percent to 89 percent, but everyone agrees that most occur extremely early in pregnancy and tend to be resorbed.[16]

The resorption of embryos is best studied in laboratory mice, but it's a common mammalian feature. Resorption occurs when the mother's immune cells break down and carry away the cells of a dead embryo, the same way they would clean up other dead cells, whether those are pathogens, symbionts, or ordinary body cells. If an embryo dies later in pregnancy and it's too large to resorb, the result is a miscarriage, or expulsion, from the mother's body. Both resorption and miscarriage are outcomes, not causes, of embryo death. So, what are the agents of misfortune? Sometimes they can be as simple as starvation. In the curious "vanishing twin" syndrome, only one embryo in a multiple pregnancy is lost. Sometimes the loss is never detected, sometimes it is revealed by ultrasound, and sometimes, rarely, at birth. Researchers theorize that it comes about as a result of competition for resources. Although mammal siblings can't eat each other, they can *out*-eat each other, by getting so many of the mother's nutrients that there isn't enough for a sibling to survive. (As we mammal parents are keenly aware, siblings can also compete for resources after birth, though only the spotted hyena is known for regular siblicide. Tellingly, this behavior appears to be triggered by extreme food limitation.[17]) Vanishing twins and other early pregnancy losses can also be caused by genetic problems, like those that lead to early chick mortality.

Hemmings, who has two children of her own, reflects, "While pregnant, I was frequently dissecting dead embryos out of eggs, and it was such a weird thing to be thinking about your own growing embryo whilst looking at dead embryos for your lab work. It was a bit surreal." When it comes to development, she knew better than anyone how frequently the process could fail. A deep understanding of the many genetic and environmental factors that can derail development was her constant companion. "I spent the majority of my pregnancy acutely aware of everything that was happening and how fragile it was. And also how amazing it was."

Developmental biologists may be uniquely predisposed to pregnancy anxiety, but they also have an awe for the process that's deepened by their intimate grasp of the details. For example, standard medical advice is to eat a healthy diet while pregnant, and it makes sense that a baby needs good nutrition to grow. But mouse experiments have proven that maternal diet can also affect a baby's *genes*. Scientists bred mice so embryos would inherit a gene from their fathers that produces obesity and yellow fur. This gene can be turned off by a compound called a methyl group, and when mouse mothers were fed methyl supplements during pregnancy, their babies were born with the gene turned off. Throughout their lives, these offspring never developed obesity and yellow fur. Of course, this specific situation doesn't exist in humans, but the process of adding methyl groups to genes, called methylation, does play a significant role in our development. Recent research suggests that the reason folic acid helps to prevent birth defects may be due to its role in methylation.[18]

We consume folic acid and other supplements with the deliberate aim of transferring them to our babies, but unfortunately, many environmental chemicals and toxins can also make their way in against our will. Their concentrations can be even higher in the embryo than in the mother.[19] Prenatal chemical exposure is a serious issue of social justice. The dangers of heavy metals discharged by industrial plants have been documented since the 1950s, but companies have been blocking research efforts and continuing to pollute already disadvantaged communities for just as long.

Even more ubiquitous than heavy metal pollutants are *endocrine disruptors*—chemicals that interfere with our endocrine system, which produces hormones. Hormones give us far more than teenage moods; they regulate and control our body's biology throughout life. Adult exposure to endocrine disruptors is typically not a huge problem, but the earlier in life we encounter and begin to accumulate them, the more opportunities they have for disruption. Endocrine disruptors fill the modern world, in plastics, pesticides, flame retardants, cosmetics, and sunscreens, and animal babies are most vulnerable to their effects—often through complex interactions over multiple generations.

The insecticide DDT (dichloro-diphenyl-trichloroethane) is one such disruptor, possibly the first to gain widespread attention for its harms. Far from a direct poison, it kills only through a cascade of environmentally mediated effects. Accumulating in the bodies of adult birds, it is broken down into another chemical called DDE, which causes female birds to produce thinner eggshells than they otherwise would. Embryos grow successfully inside these thin-shelled eggs. However, they can no longer support the weight of their parents sitting on them, leading to crushed eggs and dead chicks. Banned in the United States in 1972, DDT has lingered in the environment. Although the affected avian populations have mostly recovered, some evidence suggests that decades-old DDT is responsible for a portion of human health issues today.[20]

The drug DES (diethylstilbestrol) is another devastating disruptor. It was marketed and prescribed in the 1950s and '60s specifically to affect hormones during pregnancy—in a tragic historical irony, doctors thought it would correct "hormone imbalances" that led to miscarriage. What it accomplished instead was to seriously disturb the reproductive system of developing female embryos in utero. Women whose mothers had taken DES encountered a range of problems later in life, including increased risk of reproductive tumors and cancer. DES was banned in 1971, thankfully, but new endocrine disruptors are being invented all the time. In the early 2000s, debate erupted over BPA (bisphenol A), which is demonstrably dangerous to a developing human embryo, impacting both nervous and reproductive systems and increasing the risk of miscarriage.[21] BPA got enough bad press to spawn a plethora of "BPA-free" alternatives on the market, and to be banned from use in baby bottles in many countries (though not the United States).

One of the most disturbing things about endocrine disruptors is that their effects remain for generations after the initial exposure. When the women who had been exposed to DES as embryos had children of their own, their daughters faced the same risks, even though they had never directly experienced the drug. In mice, the fungicide vinclozolin causes reproductive issues in male embryos that are exposed to it when still in the womb, producing irregular testes, prostates, and kidneys. Even if they never see vinclozolin again, their sons suffer the same issues, and so do their grandsons. How can such an environmental effect be inherited? It seems as strange as if the pups and grandpups of the Three Blind Mice also lost their tails, with no carving knife involved. The mechanism of inheritance seems to be methylation, again. Endocrine disruptors can change what an animal bequeaths to its descendants, not by directly editing its DNA, but by altering the methyl groups attached to the DNA.[22] The twenty-first-century confluence of embryology (careful observation of how gonads develop in animal babies) and genetics (analysis of DNA to discover methylation) with environmental science (field studies on the prevalence and distribution of vinclozolin) offers a more complete picture of human effects on animal babies than we've ever had before.

Biological bilge pumps bail out babies

Humans have only been manufacturing chemicals and polluting the environment with them for a scant second of evolutionary time, but toxic chemical compounds have been a part of animal life for much longer. Some are accidental metabolic byproducts, others are produced as deliberate defenses by other animals, plants, fungi, or algae. So animals evolved tiny pumps called transporters that can remove dangerous chemicals from their cells, and these transporters are an embryo's prime line of defense against infiltration by development-disrupting contaminants. One particular chemical transporter works in mammal placentas, including those of humans, to pump out chemicals before they can pass into the embryo and cause developmental problems. However, although we know the transporter is there and we know what kind of work it does, it's been difficult to figure out how effective it is for specific chemicals. Ethical concerns limit studies on humans, and results from lab mice can be contradictory.

The most straightforward way to understand what this transporter can do is to take it away and compare results in its absence to results in its presence. So, scientists genetically engineered families of mice to stop producing the transporter. When mothers in these families are dosed with various chemicals during pregnancy, some chemicals damage the embryos that are now unprotected by placenta transporters. Embryos in ordinary lab mice without this genetic alteration don't suffer similarly. However, other chemicals affect all embryos equally, whether or not their placentas have transporters. Scientists have begun searching beyond mice to expand their understanding of this system, which will have the double benefit of greater insight into chemical risks during human pregnancy and development, and a broader view of other species' ability to cope with the same chemicals. To this end, another model system is coming to the fore.

Or rather, coming *back* to the fore, as sea urchins played a central role in the early history of developmental biology since Just used them to understand fertilization. One of the scientists who has circled back to them in a big way is Amro Hamdoun, whom I got to know when I was a graduate student at Hopkins and he was a postdoctoral researcher. Hamdoun is the kind of person who, while traveling in the Mediterranean on his honeymoon, collected sea urchins for impromptu experiments. (His spouse, Julia Cardosa, is luckily also the kind of person to get excited about this.) Hamdoun explains that urchins' ubiquity is part of their appeal for science. "You could get some species of urchin in any ocean, temperate, tropical."[23] What's more, they're prolific baby machines, producing millions of eggs on command.

Another advantage of sea urchins is their deuterostome identity. Like us vertebrates, the first indentation in their gastrulating embryos becomes an anus. Because we share a common deuterostome ancestor, we also share the majority of our genes—about 70 percent. That means many human genes of medical interest have versions that can be studied in sea urchins, including the chemical transporters in placentas. Sea urchins don't have placentas, but the same types of transporters protect their orphan embryos as they drift in the sea.

In his laboratory at Scripps Institution of Oceanography in San Diego, Hamdoun has tackled the need for a greater diversity of transporter study systems by engineering lineages of sea urchins, like mice, that don't have the transporter gene. His team accomplished this feat with a technique called CRISPR, succinctly described in *New Scientist* as "a technology that can be used to edit genes and, as such, will likely change the world."[24] Hamdoun describes it like this: "Imagine a word processor where you can't cut and paste, and then someone figures it out."[25]

When I visited Hamdoun's lab to check out his urchin-rearing facility, I found a building completely wrapped in plastic. Hand-size rips in the plastic allowed access to door handles, although all the doors were locked. When finally someone exiting the building let me in (she turned out to work with Hamdoun), I found the interior to be a normal, bustling university hall. I asked Hamdoun about the plastic. Apparently, the concrete around the hall had absorbed salt from the sea, causing the rebar to rust. This made the concrete split and crumble, he explained calmly, "creating a death hazard."

Inside, however, all was cozy. Hamdoun brought me to a small room lined with aquariums of different shapes and sizes, from large round tubs swirling with larval sea urchins to small rectangular containers holding one precious gene-edited juvenile each. Hamdoun explained that the CRISPR technique itself is easy compared to the challenges of rearing urchins through their larval stage to adulthood. "We were doing all this stuff in the literature and it was bullshit, it didn't work." Then he tackled a science project at home with his six-year-old daughter, raising killifish—a type of annual fish, with eggs that sit in the mud in diapause. They learned from a community of killifish hobbyists to rear the babies in simple eight-quart tubs, a technique that turned out to transfer perfectly to the urchin lab.

Cardosa joined us in the lab and shared some of the challenges both urchins and scientists have faced. In April 2020, a huge bloom of algae off the San Diego coast created toxic conditions in the seawater that was being pumped into the lab. Like so many tragedies, it was an indirect result of the pandemic. People weren't allowed to be in the building, so the water quality wasn't being monitored, so the algal toxins caused more damage than they otherwise would have, killing off many valuable animals.

"This was a pandemic project in every sense," adds Hamdoun. In prior years, each member of the lab would have had their own project, because they could come into the lab every day to care for their own animals and pursue their own science. But with restrictions on physical presence in the lab, "we had to redo the way we did research. We had to focus together on one thing, so we could all push it forward. This wouldn't have happened without that. Everyone in the lab has to change the water. Everyone has to care if the urchins get fed."[26] Now, as a result, the Hamdoun lab has the first animal model outside of mice that's missing the crucial transporter gene. His lab's first research on these urchins, published in 2022, was a proof of concept: the gene-edited urchins reliably produce babies and grandbabies that all grow up without the transporter, and the impact of its absence shows in their limited ability to pump out drugs and toxins. Unlike mouse development, which is all hidden inside the mother's womb, sea urchin development can be viewed in real time under a microscope, so future work can examine in far more detail the role of these chemical transporters in keeping embryos safe. Thus, an animal with no pregnancy at all may be a key to understanding our own pregnancies and their protective mechanisms.

A baby's guide to dodging the bad bugs and picking up the good ones

Long before we knew that chemicals could reach a developing human embryo, infectious disease revealed the uterus to be a mediator between embryo and environment rather than an ironclad protection. The rubella virus first alerted doctors to the fact that pathogens in the mother could affect an embryo, and far more drastically than they affect later life stages. Rubella causes German measles in children and adults, but in embryos it can produce blindness, deafness, and heart and brain defects. These days, in places where vaccines are widely available, rubella is not a great concern during pregnancy. Instead, we are warned to stay away from cat poop. Cats can carry the parasite *Toxoplasma gondii*, which often produces no symptoms in adults but has a chance of causing eye and brain damage when contracted in the womb. (My husband took over litter box duty for our two cats while I incubated our offspring.)

With all our concern over healthy pregnancy and birth, we rarely think of celebrating the important symbionts that we also "catch" from our mothers. Mice need to pick up the correct gut microbes at birth in order for the blood vessels in their guts to grow properly, and these gut microbes even influence brain development.[27] Microbes from our mothers are also, perhaps surprisingly, crucial to the development of our immune system. Once thought to be a line of defense primed to attack anything that isn't the self, the immune system is increasingly described as a collaboration between host and symbionts. As Gilbert puts it, "The immune system is immature without the microbes."[28]

Various studies suggest that microbes can colonize developing embryos before birth. Some scientists have found evidence for microbes in the amniotic fluid itself, though others remain skeptical. Mouse experiments show that bacteria fed to a pregnant mother can indeed be transmitted to her pups, perhaps in a process mediated by the mother's own immune system. However, the vast majority of our body's bacterial colonization begins at birth, when the amniotic sac ruptures and we encounter an outside world teeming with tiny life. The microbes acquired by passing through the birth canal are the most beneficial, leaving babies born by Cesarean section at a disadvantage. Recent studies have found restorative effects of dosing C-section newborns with "probiotics" consisting of the mother's vaginal secretions or feces.[29] (Fecal bacteria are routinely encountered during vaginal birth, for reasons that I hope you can imagine without needing me to go into great detail. This interaction recalls the pap of koalas and the pedestal of dung beetles.) After twice experiencing the raw mess of vaginal birth (three times, technically, but I don't remember being the one extruded), I'm glad to know that it served the purpose of introducing my children to appropriate microbial partners. Or perhaps I should say, introducing my own intimate microbial partners, with whom I've had a lifelong association, to a couple of brand-new hosts.

Within a few minutes, each baby in turn latched on to a nipple, and started a whole new process of nutritional and microbial transfer.

Feeding from your mother doesn't stop after birth. Lactation, nursing, breastfeeding, whatever you want to call it—mammals engage in matrotrophy for days to years of our early life. And although we placental mammals drink milk from nipples, that's not the only way. Echidnas and platypuses, the only living members of the once-larger egg-laying mammalian group called monotremes, simply lap up milk that their mothers sweat out of pores in their skin. As if laying eggs wasn't weird enough.

Seriously, though, the fact that monotremes both lay eggs and sweat milk gives us a window into the evolutionary origin of lactation. Scientists theorize that it began as a way to hydrate eggs. Remember the frogs who pee on their broods to keep them from drying out? Early mammals probably sweated on their eggs to achieve the same end. Over time, those who added nutrients to the sweat for their embryos to absorb became increasingly successful, until that specialized sweat evolved into milk, a substance with the dual purpose of nutrition and hydration. Hydration is still an important purpose of lactation today—human infants don't usually start to drink water until they're at least six months old.

Mammal infants are no mere passive receivers of milk. Many features of mammal babies are adaptations specifically to get their main source of nutrition. Newborn humans can crawl up their mother's body to find a nipple, an ability they will later lose and take months to regain. They

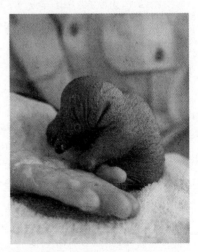

Baby platypuses, as well as baby echidnas like the one pictured here, were dubbed "puggles" when a researcher noted their resemblance to plush toys of the same name.

have abundant and broadly distributed taste buds, from tongue to tonsils, and they show an innate preference for the sweetness of breast milk over other flavors. Their suckling creates a dialogue with the mother's body, influencing the nutritional content of the milk as they grow. Marsupial babies like kangaroos may share their pouch with a much older or younger sibling, but they don't share nipples—each baby has its own teat, which produces age-appropriate milk. Baby mammal mouths are shaped to suck, their paws stimulate milk production, and as soon as they're mobile, they can zero in on sources of milk, even stealing it from unrelated mothers. Milk theft is common in sea lions, seals, reindeer, camels, and giraffes, although some cases may be less of a theft and more of a consensual interaction. An overfull teat is a distinct discomfort, and emptying it into a willing mouth is a relief. Mother bats even seek out unrelated babies to feed if their own children's tummies are full, since they need to get rid of the milk weight before flying.[30]

It's probably no surprise to hear that milk carries beneficial microbes, as well as food for microbes. While my babies nursed, they were nourishing and structuring the communities in their new little guts. These helpful symbionts not only aid with digestion but can also protect against disease. The bacteria that infant monkeys obtain from their mothers' milk summon a particular kind of white blood cell into the gut, which then fights off pathogens, including *Salmonella*.[31] Research on this topic has accelerated in recent years, and I expect that by the time you read this book, new relationships will have been uncovered among microbes, milk, and babies.

Lactation is undoubtedly a female thing—but that doesn't mean it's never a male thing. Fathers of at least two species of bat have been observed to produce milk and nurse their young, though not as much as females. Both male and female birds can produce crop milk, depending on the species. Crop milk is distinct from mammal milk in that it isn't produced in mammary glands and extruded through fancy pores but rather produced in the digestive system and regurgitated like food. However, it is produced by the parent's body specifically as baby food, and the nutritional content is surprisingly similar to that of mammalian milk. More "milks" are likely to be found in non-mammals in the future—spider milk was discovered for the first time in 2018.

Even the nonliquid baby food produced by the curious legless amphibians known as caecilians is similar in nutritional profile to mammalian milk. Different species of caecilians appear to have evolved on a matrotrophic continuum between chewing on mom from the outside and chewing on mom from the inside. Some species lay eggs that hatch larvae with special teeth quite different from those of adults, perfectly suited to peeling off the outer layer of their mother's skin, which is enriched with fats and nutrients for this purpose.[32] Other caecilian species can get pregnant, and inside their bodies the babies develop the same unusual teeth, which are used to scrape nutrition off the inner walls that surround them.

Some frog parents have similarly used their bodies to provide all the food their babies need. Tadpoles of Rabbs' fringe-limbed tree frog, a species now likely extinct, got their early nutrition by eating their father's skin cells. At least one Australian sea urchin, of all curiosities, may engage in a similar practice. Marine biologist Richard Emlet observed these urchin mothers brooding eggs on their back. "You have one egg anchored in place by three little mucus straps," he says. "And then it hatches and it has tube feet that it can hang on to Mom with. It hatches with no spines, it's just this little muffin with tube feet. And the spots it sits on Mom are these red, raw sores. And I wonder if it eats Mom."[33]

Earwigs, like physogastric mites, make the ultimate sacrifice for their children. These big crawlers tend to wig people out, with their pincers and their propensity to squiggle unexpectedly from under rocks or behind furniture, but they can't hurt you, and once you know what they do for their babies, you might find yourself a little kindlier inclined toward them. In many species, a mother earwig protects her eggs until hatching, brings food to her babies, and finally lets them consume her body.[34]

It might seem that a deep commitment to the care and feeding of offspring would put them in a better position for survival, compared to other species that simply release their babies into the wild. That may be true for individuals whose parents have the resources to nourish them, but there's also a hidden cost to this close relationship between parent and child. If anything happens to the parents or to the habitat into which they've brought their children, options for recovery are limited. Capricious environments can wreak havoc on a tightly knit life cycle but offer an advantage to species with independent, adventurous larvae.

It's time to take a look at the world's myriad larval forms, and at what happens when they strike out on their own.

PART II

——

SALAD DAYS

UNACCOMPANIED MINORS

Where Do the Escargot?

> We are the seeds of the tenacious plant, and it is in
> our ripeness and our fullness of heart that we
> are given to the wind and are scattered.
>
> —Kahlil Gibran, "The Farewell"[1]

Linguistically, "semen" and "seed" are the same, but biologically, it's the eggs of animals that are analogous to the seeds of plants. Both egg and seed contain an embryo, along with nutrition for the embryo to grow. Both have a protective covering and, in many cases, a means for dispersal. They are tiny, efficient, hopeful. They are capsules sent out into time and space, connecting *here and now* to *there and then*.

The act of dispersal is how life sets up shop in new places. It's different from migration, in which animals like monarch butterflies travel regularly between established homes. A migrating butterfly wings her way from Indiana to Mexico, following a route taken by generations of her forebears. A dispersing butterfly gets blown out to sea by the wind and might be lucky enough to land on an island where she can lay her eggs. Dispersal is always happening, from butterflies and dandelions in breezes to squid and spores in ocean currents. And although adults can certainly disperse, much of our environment is shaped by travelers too tiny to see. Sometimes for better and sometimes for worse—both wildflowers and weeds show up wherever their minuscule seeds have been carried. We even use the word *weedy* to describe animal species that easily colonize and flourish in new habitats.

Most often, dispersal doesn't "take." We don't know exactly how often, because it's impossible to catalog every failed dispersal event—every butterfly that was blown off course and died, every seed that landed on concrete and never germinated. However, we do know that human activity has vastly increased the opportunities for dispersal. Animal parents lay their eggs in seemingly secure locations, only to have them carried away on the bottoms of shoes or boats, or packed with produce or lumber and shipped around the world. Seeds are circulated from port to port on cruise ships and jet airliners. Not every organism that arrives in a new habitat can make it their home, but those who do have the capacity to wreak havoc. As they eat or infect local species that had no chance to evolve defenses, these invasive species can collapse ecosystems. We know that such invasions must have happened from time to time long before we humans came on the scene. After all, any animal or plant lives where it does today because an ancestor dispersed to that location. But the speed and frequency with which it's happening now is unprecedented.

Humans do everything a little too enthusiastically. As a species, we remind me of toddlers. We're still in the process of learning to regulate ourselves. We overwater the garden, we squeeze the cat too tightly, we eat all the cookies instead of saving some for tomorrow. And we ship our food, our tools, our toys, and our bodies all around the world, without taking a moment to consider what else we might be unintentionally moving around.

That's how the waterways of North America got infested with European zebra mussels, for example, which outcompete native species and clog water intakes. Zebra mussels can only survive in fresh water, so we know that they didn't disperse as adults attached to the outside of ships. They would have died in the salty Atlantic. Instead, it's likely that their free-swimming larvae, microscopic and transparent, made the crossing inside the freshwater ballast tanks of ships. Ballast is a stabilizer, water weight that ships take on in port to keep them from tipping during the voyage. The amount of ballast needed varies by cargo load, so ballast is often slurped up in one port and discharged in another—along with everything living in that water.[2]

One of our fundamental intuitions about life is that animals can move around, while plants are rooted in place. Given their immobility, we can understand why plants need a specific "dispersal phase" of seeds that can surf the wind, hitchhike on bear fur, or ride inside bird guts. The case of the mussels illustrates the fact that many animals, especially aquatic animals, are surprisingly similar to plants in this way. Animals that are *sessile* (permanently stuck in one place) or *sedentary* (disinclined to move around) have developed an innovative array of highly mobile larval forms. These larvae are planktonic, meaning that they drift freely with the waves and currents. The adults that we see when exploring a pond, a tide pool, or a coral reef are the successful settlers, the products of a perilous peripatetic phase.

Richard Emlet, the marine biologist who spotted sea urchin babies chewing on their mother, began his career in research by waking with the low tide, no matter what time of day or night it happened to be. He then took a small boat fifteen miles off the coast of Panama to a rocky island, which offered no relief from the tropical sun when it rose. He was there to identify the adult inhabitants, like barnacles, crabs, and fish, but he realized that the makeup of the entire ecosystem was dictated by which animals arrived as babies from the plankton.

"That meant larvae, larvae, larvae," he says. "There was nothing else in the world worth studying."[3]

Plant parallels

Fu-Shiang Chia was widely known for his work in developmental biology, but he was also a poet. He embodied a spirit of connectivity across space and time and the flexibility of making unexpected linkages, just like dispersing larvae. In his long life, from 1931 to 2011, he lived in rural Shandong, China; Taipei, Taiwan; Seattle, USA; and Edmonton, Canada. He published hundreds of research papers, as well as books of original poetry, and a translation of the world's oldest lyric poetry collection ("Airs of the States" from the *Shi Jing*) into both English and modern Chinese.

At Friday Harbor, Chia studied larval settlement, the momentous event in which an animal leaves its planktonic lifestyle to select an adult habitat. Building on earlier work, he discovered that eleven-day-old larval anemones would settle on worm tubes, but if they were given only plain dishes, they wouldn't settle until they were twenty-seven days old. These findings, in turn, anticipated current cutting-edge research on microbe-mediated settlement (see chapter 9).[4] One of his students also showed that post-larval snails could use mucus threads to drift with the current, extending the possibility of dispersal for these organisms from planktonic larvae into the young-adult stage. This behavior is analogous to that of the spider babies in the children's classic *Charlotte's Web*. These tiny spiderlings engaged in ballooning, making silk threads to carry them far afield on the wind. Several kinds of caterpillars also use silk for wind travel, dangling from trees to be whisked off on a breeze.

The caterpillars of European spongy moths take their wind transportation game above and beyond. Brought to America by humans hoping to start a silk industry in the 1800s, they escaped cultivation and began to breed in the wild. These caterpillars eat so many leaves from such a diversity of plants that they became a serious pest, and their ability to spread has made them difficult to control. Eggs are often laid on cars or trucks and transported long distances. People moving across spongy moth quarantine lines are required to inspect all outdoor items, from toys and bikes to lawn chairs and rakes, for spongy moth eggs. These measures can help slow the species' spread, but it's hard to do anything about the fact that spongy moth caterpillars grow bristly hairs full of air pockets. This makes them so light that winds can carry them as high as 2,000 feet (610 m) and as far as 5 miles (8 km) a day.[5] They're the animal version of dandelion seeds.

The dispersal of plant seeds is incredibly diverse, from fluff that catches the wind to burrs that catch on your sock, from fruit seeds that pass through a mammal's gut and get deposited in a pile of fertilizer to parasitic mistletoe seeds that pass through a bird's gut and get wiped off on the branches of a new host tree. Thus, although individual plants can't pick up roots and take a walk like the tree-people Ents in *The Lord of the Rings*, plant populations and species can "move" by seed dispersal. This is advantageous for the individual who lands in a new habitat and finds abundant resources; it's also advantageous for the species, which is more likely to persist when environmental conditions change. These same advantages hold for dispersing animals.

And seeds, like baby animals, are embryos—not a single unfertilized egg or sperm cell, but a multicellular organism. Their multicellularity allows both plant and animal embryos to grow complex structures for dispersal. Plant embryos in seeds are arrested, developing to a certain point and then pausing until the conditions are right to continue. Several animals produce similarly arrested embryos or larvae, like the diapause eggs of annual fish or the dauer larvae of roundworms. Pausing development for an indeterminate time offers flexibility in a fluctuating environment. Both plants and animals often tie their reproduction to the seasons, producing babies only at particular times of the year. If those babies can then put their development on hold, they can make use of year-to-year variation in temperature, sunlight, and availability of food.

Many eggs, especially fish eggs, are packed with fatty yolk that makes them float, carrying them up from the seafloor into the ocean currents. Yolk is often consumed both before and after hatching, so freely drifting larvae can still contain a considerable amount of it. As they grow and use it up, the composition of their bodies changes. They become heavier than the surrounding water, sinking back down to the bottom. Sinking is an important behavior, even though it is the default for all the marine larvae that must return to sand or rocks for their adult lives. Larvae sink so much that at least one predator, the comb jelly, may have evolved to take advantage of it. These transparent and voracious jellies have been observed swimming upside down with their mouths wide, an effective approach for collecting sinking snail and clam larvae—meals that are commonly found in their stomachs.[6]

Useful as they are, floating and sinking are far from the only motions in larval repertoires. Cephalopod hatchlings swim by jet propulsion; fish larvae swim by undulation or paddling. Perhaps most efficient of all is to hitch a ride on a larger, stronger swimmer. The unioid clams are one of the best examples. They live in fast-flowing streams, and their babies grow a special shell shaped like a bear trap that can hook on to fish gills. Adult brood pouches have adapted to look like live minnows, tempting predatory fish closer. When a large fish is near enough, the parent releases its larvae, which then clasp on to the fish for a free ride. These piscine taxis are strong enough to swim both up- and downstream, dispersing clam larvae hither and yon.[7]

Water dwellers are even capable of catching a lift from aquatic birds. The appearance of invertebrates and fish in isolated ponds and lakes long invited speculation that they arrived on flying feet or feathers, but no one could ever prove it. Recent research has illuminated a different and surprising mechanism for bird-mediated dispersal—poop. Amazingly, although most larvae and eggs eaten by birds are doomed to digestion, a small but relevant percentage

To trick host fish into carrying their larvae, freshwater mussels use elaborate lures that can resemble minnows (as shown here), worms, or crayfish.

can survive the trip through the digestive tract to emerge still living and growing out the other end.[8]

On land, walking-stick insects have evolved eggs that mimic delicious-looking seeds, and one study has shown that these eggs, too, can survive being eaten and defecated by a bird. Even more well established is the fact that ants collect both seeds and seed-mimicking walking-stick eggs. Brought back to the ant colony and hidden underground, the developing walking-stick babies are protected from predators and parasites (see insert, photo 9). In this case, dispersal may be a secondary benefit or even irrelevant; the real advantage is safety.[9]

Volcano hopping

Lauren Mullineaux was brought to larvae by her interest in seed dispersal. She had attended college near a desert, where she began to wonder about the dispersal of desert plants between oases. When she had the opportunity to join a deep-sea research cruise, the similarities between these two environments caught her attention. Deserts and the deep sea are both wide expanses of limited resources, dotted with small localized sites where water (in deserts) or food (in the deep sea) is more abundant.[10] How do animals find such havens, and travel between them?

Until the second half of the nineteenth century, the prevailing scientific view was that nothing could live in the deep sea. It was too deep, too cold, too dark. Although a few counterexamples of life had been raised incidentally by deep dredges, the "lifeless deep" paradigm wasn't overturned until an expedition in 1868–70 collected hundreds of new species from depths well over 3,000 feet (914 m).[11] Even then, our revised view of the deep sea was of a bland, homogeneous environment. The animals that lived there wouldn't have any use for dispersal, scientists thought, because there was nowhere to go that was any different from anywhere else. Then, in 1977, we learned about hydrothermal vents.

These are locations where the earth's crust cracks open to release life-supporting chemicals and intense heat. We've already met a few species that use the warmer water as a nursery for developing embryos, but many others evolved specifically to live off the sulfur and methane leaking from the bowels of the earth. These animals depend on volcanic activity for their survival, but volcanoes don't live forever. A hydrothermal vent lasts a few years—maybe five, maybe twenty. Eventually the roiling release settles down, the site cools off, and the only species that will survive are those with the capacity to disperse to a new vent.

Mullineaux, now at Woods Hole Oceanographic Institution, found that many of these animals produce large, yolky, buoyant eggs, which could easily distance themselves from the seafloor and be carried on ocean currents. Furthermore, some species enter developmental arrest at the two-cell stage, facilitating even more extensive dispersal.[12] But how far do they actually go?

One of the most abundant vent animals, and the first one that I learned about, is the giant tube worm, which looks like an enormous tube of lipstick. These fascinating worms grow in groups so dense they look like fields of grass. They have no guts and depend entirely for nutrition on symbiotic bacteria that metabolize vent chemicals. The large red plume that waves above their white tube is a teeming bacterial farm.

Vent tube worms look so strange that they were thought at first to constitute their own phylum—the highest-level grouping of animals. For context, everything with a backbone, from fish and frogs to hens and humans, belongs to the same phylum, Chordata. It might sound strange to erect a whole new phylum for a new kind of worm, but in fact there are already at least half a dozen worm phyla (the plural of *phylum*).

The word "worm" makes most people think of earthworms. Or, if you've taken a puppy to the vet or worked in medicine, maybe you think of parasitic worms: roundworms and tapeworms. These three kinds of worm are different enough to be placed in three entirely different phyla, and still more worm phyla partition off the ribbon worms, arrow worms, peanut worms, and penis worms. All these common names are derived from visual similarities with the titular objects, though I have to admit that penis worms resemble phalluses far more than peanut worms resemble legumes. (Inchworms, mealworms, and wax worms are actually larval stages of insects, and thus belong to the phylum Arthropoda.) Why do we have so many phyla for animals that are all long and thin and limbless and squishy? Well, this superficial similarity of form hides extraordinary differences, none more remarkable than the life cycles and larval stages of each group.

Indeed, it was through their larvae that vent worms were eventually identified as belonging to one of these existing worm phyla. Deep-sea biologist Craig Young collected their eggs and watched them develop into the distinctive larval form called a trochophore. These larvae look like spinning tops, wearing a skirt of hairlike projections called cilia and sprouting another tuft of cilia like a topknot. The trochophores allowed Young to assign vent worms to the same phylum as earthworms and bristleworms.[13] (Humorously, Young and his colleagues had been assiduously collecting both eggs and sperm from the vent worms, mixing them in the lab, then congratulating themselves on successful fertilization. Years later, they realized that the eggs they were collecting had already been fertilized before collection.)

Young and Mullineaux worked together to rear baby vent worms for the first time in a laboratory, using pressurized aquariums. They found that embryos developed in thirty-four days to become swimming larvae but never developed mouths. Since they are unable to eat, the life span of these

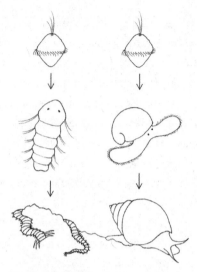

Similar-looking trochophore larvae reveal drastic differences as they grow. Some develop into setigers and then adult worms (left); others develop into veligers and then adult snails (right). Not to scale.

babies must be limited by their yolk reserves. The scientists measured their metabolism to find out how fast they would burn through their parent-packed fuel. These experiments provided a range of thirty-four to forty-four days during which the larvae could disperse. If they had not found adult habitat in that time period, they would die.

Data on the flow of ocean currents above and around hydrothermal vents could then be used to model all the possible travel routes of these larvae. The researchers found that frequent fluctuations and reversals of current in the area were incredibly important in determining how far the larvae went and whether they were more likely to return to their parent's vent or to colonize a new site. The drifting larvae produced by tube worms face, if you'll excuse the expression, a sea of possibility. A significant number are predicted to settle within tens of miles, close enough to return to their vent of birth, which explains how a single vent can become so densely populated. A smaller but still relevant number could be carried along the ridge of volcanic activity where more vents are forming, seeding new populations at new vents. And a few long-lived larvae could be carried off the ridge entirely—often a death sentence, but from time to time a lucky break, leading to the colonization of a previously uninhabited site.[14]

Dispersal and immortality

Instead of "founding fathers," a great many animal populations have "founding babies." Starfish and mussels in a tide pool, barnacles and worms on the bottom of a boat, and spiders in a barnyard all arrived in the exuberance of their youth. We know that animal babies can reshape ecosystems by establishing new populations where none existed before, but we're still figuring out the scale on which this typically occurs.

Rudi Scheltema, "one of the grandfathers of larval biology,"[15] was the first to measure the possibility for extreme dispersal by identifying larvae from nearshore organisms far out to sea in the mid-Atlantic. With the potential established, many questions remained about the implications of these remote larvae. Where exactly had they come from, and could they mature to adulthood or were they doomed by the distance they'd traveled? Ocean-wide currents are more complex and challenging to model than the smaller-scale currents between deep-sea vents, and techniques for proving a particular larva's point of origin took many years to develop. Over time, a combination of physics, chemistry, and genetics started pinning down the answers.

When Rachel Collin arrived as an undergraduate student at Woods Hole, she'd planned to work with Mullineaux. However, Mullineaux was on maternity leave—busy with human developmental biology, as it were. "So they put me in the lab of Rudi

Scheltema, who was about eighty at the time and retired and was like, 'Hmm, well, why'd they stick this student here?'" says Collin, laughing. But Scheltema was still organizing research cruises, because being "retired" in academia doesn't mean that you stop researching. The project Collin joined was an investigation of whether far-out larvae could metamorphose if they encountered the right substrate. Unfortunately, the cruises were challenged with bad weather, and trying to collect plankton in the midst of a storm turned out to be a bust.[16]

Undeterred by early fieldwork mishaps, Collin went on to graduate school at Friday Harbor and made a career of studying reproduction and development. Now at the Smithsonian Tropical Research Institute in Panama, she discovered that some of the sea star larvae in the mid-Atlantic can travel halfway around the world *and* successfully establish new populations. The evidence came from larvae that had an extra trick up their sleeves: asexual reproduction, or budding off clones. Within the echinoderms, a phylum that includes sea stars and sea urchins, larval cloning is fairly common and can be stimulated by everything from predators[17] to abundant food.[18] You could say they've found a way to both have and eat their reproductive cake. Over time, they reap the rewards of both sexual reproduction (genetic shuffling for greater diversity) and asexual reproduction (rapid multiplication without needing to find a partner). Embryonic cloning in parasitic wasps and parthenogenesis in reptiles and birds show that many different animal groups benefit from creatively combining reproductive strategies.

Cloning sea star larvae have been found in the Atlantic and Caribbean since 1989, but no one knew which species they belonged to. Collin solved the mystery with genetics. Simply sequencing the DNA of an unknown organism doesn't automatically tell you what species it is; you need to find a match with something known. So Collin plugged the DNA of the cloning larvae into a genetic database, and found a match from a species that had been sequenced by Gustav Paulay. "He's a way-back-in-the-day Friday Harbor guy, too," Collin tells me. "I mean, *way* before me." She contacted Gustav and found out that he'd collected the DNA from adult sea stars in Moorea, an island in the South Pacific, over 5,000 miles (8,047 km) from Collin's Caribbean samples. That journey would be epic enough if the larvae were taking a straight shot through the Panama Canal, but the current patterns indicate that they aren't—they're traveling around Cape Horn instead, a journey of at least twice the distance. These larvae may be able to cover the miles by replicating themselves en route, extending the time they can spend in the water with multiple "generations" of babies. Although no one has yet discovered adults of this particular species in the Caribbean, a different sea star that also has cloning larvae has been found as an adult in both the Indo-Pacific and the Caribbean, supporting the idea that such long-distance transport can establish new populations.[19]

Researchers typically calculate how far and in what directions larvae will drift based on current patterns and a number called the pelagic larval duration, or PLD. This number is calculated for a given species by testing how long its larvae take to metamorphose in the laboratory—like the vent tube worms that were found to survive for about forty days on their yolk reserves. Researchers then create mathematical models of dispersal, assuming that any larvae who reach the end of their PLD without finding a good location to settle are, I'm sorry to say, dead in the water.

However, this reliance on PLD might be misleading, and not just because cloning might reset it. In the wild, larvae might encounter conditions drastically different from the usual lab setup. There's evidence that some species can stay in the plankton almost indefinitely, waiting for the right place to settle. The developmental biologist Amro Hamdoun, who created sea urchins without toxin transporters, also tried a second "pandemic project" of raising urchin larvae without solid food. He calls it a "weird thing that we'd never try under normal circumstances, of course expecting that it wouldn't work at all, and it did."[20] Normally, these sea urchin larvae settle down and metamorphose into adults within three weeks. By providing them only with dissolved nutrients that they could absorb from the water through their skin, Hamdoun was able to keep them swimming as plankton for more than twice that long. They had shrunken guts and never metamorphosed, but it's possible that, in the right conditions, they could regrow their guts and continue developing.

That's what Amy Moran, a biologist at the University of Hawai'i, accidentally discovered. She tells me a story about a large batch of sea urchin larvae that she fed once and then left in the cold room, a refrigerated laboratory space. They were surplus, not the subjects of any particular experiments, and for several weeks she forgot about them. When she checked and found them, to her surprise, still alive, she gave them a second round of food. The next time she came back to check on them, they had metamorphosed to tiny juvenile sea urchins. "I fed them *twice*," she says incredulously. "I've been thinking about that ever since. That is so *not* how you're supposed to rear sea urchin larvae."[21] But how often does nature follow lab protocols? Could larvae survive far longer and in far rougher conditions than typical PLD experiments suggest? Moran's experience with the urchins, and similar results with oysters, have motivated her to start a new research project on the flexibility of larvae in their natural environment. "I think they're going to be able to stay in the plankton long enough to go anywhere. The question will be how often they do that."[22]

Temperature and babies: flexible or finicky?

Some sea creatures, like salmon and lobster, have been hunted heavily by humans since long before we invented departments of fish and game and spreadsheets to track our haul. Others, like Kellet's whelk, may have been occasionally collected but only recently drew the attention of substantial fishing operations. (It's no coincidence that new fisheries arise as old ones crash. Ever since we began overharvesting prey, we've had to turn from one species to the next to the next.) Kellet's whelk is a sea snail on the California coast, with a pretty spiral shell that can grow as long as your face from chin to forehead and a meaty foot that can be cooked and eaten in soups, salads, spaghetti, and more. In the final decades of the twentieth century, people began to catch more and more of these whelks, drawing the attention of managers and scientists who wondered if this slow-growing species could handle the heightened harvest. At the same time, Kellet's whelks began to show up north of their previous range limit. They had been familiar occupants of the shoreline from Baja California to Santa Barbara, and now they could be found all the way up to Monterey—an expansion of 200 miles (322 km). Because there's an elbow of land just north of Santa Barbara called Point Conception where the ocean becomes significantly cooler, the whelk's range change was initially assumed to be caused by a change in sea temperature. If the water between Santa Barbara and Monterey had warmed, then previously inhospitable habitat might be welcoming the whelks. But biologist Danielle Zacherl knew the evidence for temperature effects was limited, and she wanted to find out if larval dispersal could offer more insight.

Curiously enough, marine snails beget the same larval form as marine worms: a spinning, tufted trochophore. They quickly develop into a unique form of their own, however, which looks like a baby snail with an enormous veil. *Velum* is the Latin word for "veil," hence these larvae have been christened *veligers*, or veil-bearers. The sharp-eyed embryologist and whimsical poet Walter Garstang wrote in 1928:

> The Veliger's a lively tar, the liveliest afloat,
> A whirling wheel on either side propels his little boat;
> But when the danger signal warns his bustling submarine,
> He stops the engine, shuts the port, and drops below unseen.[23]

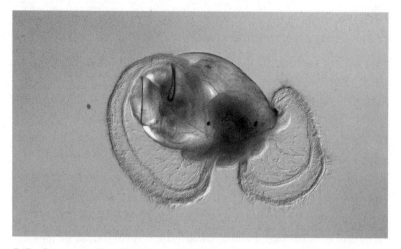

Like the sea snail it will become, this veliger larva lives in a shell and peers out with a pair of eyes. Unlike its adult form, it uses two winglike lobes lined with bands of cilia to swim and collect food.

The "whirling wheel" is the velum, which is lined with cilia like those that encircle the trochophore. Cilia beat together like little oars to generate a water current, which can collect tiny particles of food and also provide locomotory power. The veliger's "little boat" is its shell, typically transparent, which will become the core of the adult shell the snail continues to build throughout its lifetime. And the behavior of whisking the velum inside the shell is a reaction that may allow the larva to sink out of reach of certain predators (though not, perhaps, the voracious comb jellies).

Because they are small animals that can be kept in a laboratory under diverse experimental conditions, veligers are one of the "lab rats" most useful in the study of ocean acidification. This is a process occurring alongside climate change, and driven by the same excess carbon dioxide in our atmosphere. As the ocean absorbs some of this carbon dioxide, the water becomes more acidic, but long before it's acidic enough to erode adult seashells, it begins giving babies a hard time. Thus, veligers and other larval forms that produce skeletons can serve as a bellwether.

The hard parts of veligers are not only informative for studies of environmental change but also for calculations of larval dispersal distance. This was the key to Zacherl's research on whelks. She could collect plenty of whelk larvae from the wild and plot the locations of their capture on a map. But how to figure out where each veliger had started its journey? Their hard parts integrate chemicals from the seawater around them as they grow, embedding a chemical signature that can be matched to the chemical signatures of different water masses. Larvae carry a "fingerprint" of their natal location.

I say "hard parts" instead of "shell," because the particular pieces most useful for chemical fingerprinting are the animal's tiny ear bones, called statoliths. These microscopic lumps of calcium help the snails tell up from down, quite similar to the otoliths in your inner ear that keep your balance. Statoliths grow as the animal does, laid down in layers like tree rings, so the chemistry of the innermost layers can tell scientists about the animal's earliest environment. After overcoming the frustration of extracting miniature statoliths from baby snails, Zacherl was able to figure out where they were born, gaining an accurate picture of how far they'd traveled.

Her results indicated that the expanded range of the species was intimately linked to the dispersal of their babies. The water currents around Point Conception usually prevented whelk veligers from crossing into northern waters, but changes in these patterns due to El Niño events could periodically bring larvae north of the point. At the same time, warming water due to climate change might increase their survival as both larvae and juveniles. It was the impact of changes in both temperature and ocean currents on the early life stages, not the adults, that most likely caused the range expansion.[24] (Zacherl kindly allowed me to borrow her study title "Where Do the Escargot?" for this chapter.)

In some cases, larval dispersal can play a major role in restoring damaged environments. The Great Barrier Reef off the coast of

Australia suffered two major coral bleaching events, in 2016 and 2017. Heat waves caused many corals to throw out their algal symbionts, losing their color and, shortly afterward, their lives. Bleaching itself doesn't kill corals, but because they depend on their algae for food, they often die of starvation. This happened in many parts of the reef, including the remote resort of Lizard Island. However, by 2021, Lizard Island's entire reef was regrowing, seeded with larvae that had traveled from farther afield. The recovery offered an encouraging note after a bleak symphony of devastation.[25]

Sadly, not all larvae can adapt to significant shifts in their environment. Scientists have found worrying implications of climate change for the dispersal of snail larvae on the Atlantic coast of the United States. Unlike the Kellet's whelks on the Pacific side, these Atlantic snails prefer lower temperatures. And yet distributions of many species have been shrinking, retracting from the cooler depths toward the warming shallows. Developmental biologists and physical oceanographers tackled the problem together, analyzing both living and nonliving aspects of the system. They figured out that climate change has led parent snails to release their eggs earlier in the season, when south-flowing currents are stronger. Larvae that would otherwise have been able to settle and populate the cool depths are now being carried toward tropical waters and away from good homes.[26]

As for species that are native to the tropics, Collin has found that early life stages are extra sensitive to rising temperature. She compared the embryos and adults of eight Caribbean sea urchin species and found that embryos were universally more susceptible to damage from high temperatures. The geographical ranges of these species are determined by the tolerances of their embryos; the coolest parts of each range coincide with the lowest temperatures at which that species' development is successful. As for heat? Where the water is warmest, the babies are already at their upper limits. If a heat wave strikes, they're gone.[27]

The responses of larvae to their environment can expand or contract the range of a species. Larvae can be heroic rescuers, rebuilding a site after death and destruction—or they can be unwelcome party crashers, *causing* death and destruction by invading an unprepared ecosystem. The curious blend of flexibility and fragility during this life stage can tip the balance between a species' survival and its extinction.

When larvae are bold adventurers whirling from sea to sea and coast to coast, it's easy to see how they can have such an outsize impact. However, even relatively stationary babies, like caterpillars munching away at the roadside, form critical links between their own species and many others, including humans.

IT'S JUST A PHASE

Why Babies Look Like Aliens

Under this loop of honeysuckle,
A hungry, hairy caterpillar,
I crawl on my high and swinging seat,
And eat, eat, eat—as one ought to eat.

—Robert Graves, "The Caterpillar"[1]

In rare cases, young animals resemble miniature adults in form and behavior. Foals and ducklings can walk and swim, respectively, within hours. But most newborns both look and function very differently from their parents. This is a critical connection—babies look different *because* they must function differently, or, put another way, baby interactions with baby environments have favored different adaptations at this life stage. Altricial birds are born blind and featherless, looking like wet rats rather than hawks or herons. But they aren't simply incomplete adults; they have their own complex features, such as colorful gapes and carefully tuned peeps. Baby mammals are often blind as well, but marsupial joeys have strong arms to climb into their mother's pouch, and meerkat pups make precise begging calls that attract the attention of every meerkat mom in the vicinity. Langur monkeys are eye-catching orange at birth, only changing to the mature black-and-white of adults after they've gotten the intense maternal attention they need.

Some differences in form are due to simple size limitation. Adult body types often don't work as well when they're shrunk down, as I found for squid paralarvae. The word *paralarvae* was coined for baby squid and octopuses because their early life stages are notably different from adults, with distinct body shapes, fins, and tentacles—but not *as* different as a true larva like a caterpillar is from a butterfly. Paralarval squid swim with jet propulsion, like adults, but because fluids operate differently at tiny scales (for example, an ant actually can carry a droplet of water), they need a much larger opening relative to their body size to successfully squeeze the water out. Thus, in addition to their curious fins and tentacles, they have an enormous funnel, which they adjust while swimming to maximize their efficiency.[2]

Still, paralarval squid are recognizable as squid, like human babies are recognizable as humans. Consider by contrast that baby butterflies look like worms, baby worms look like spinning tops, and baby starfish look like teddy bears with tentacles. Larval forms like these are so weird that many of them were initially described as separate species from their parents. Even today, most of us haven't seen a fraction of the larval diversity the world has to offer. The *why* behind the weirdness isn't that Mother Nature was feeling whimsical and took it out on her babies. Every strange feature is a response to the environment, every spine and tentacle an adaptation to the small sizes and challenging habitats of childhood.

Our default mental picture for any animal is nearly always its adult form. Let's find out why perhaps it should be the larval form instead—or at least the two side by side.

"They are the main game"

Life has costs. We living things need a certain amount of fuel, from the gases we breathe in to the sugars we break down, just to get through the day. And we need additional fuel for special tasks, whether we're mother crabs flapping our abdomens or larval sea snails making our first shells. Growing up, in fact, is expensive. Australian researcher Dustin Marshall has calculated the cost of development at different temperatures for a wide range of animals, and he found to his surprise that most species are already developing at their optimal temperatures. A little warmer or a little colder, and the process gets far more expensive. "They're on a knife edge," he says, with temperature deviation in either direction making their development far more costly.[3] That's because the warmer the environment, the faster babies can develop—but also, the greater their metabolic needs for food and oxygen.

I ask him about the resilience of these babies. Does this narrow tolerance reveal how vulnerable they are, or does it illustrate instead how well adapted they are to their niches? Both, says Marshall. "On the one hand, they're little eggs and they're terrible at things and many of them die. That's definitely true. At the same time, selection is never going to be more efficacious than for that life history stage." Natural selection, the driving force behind evolution, acts most strongly on the youngest life-forms. Genes that make you more likely to survive until you reproduce will be kept in the population, whereas genes that hurt your chances will be weeded out. The results of this selection aren't always obvious, however. As Marshall points out, "We tend to look at, say, an eagle and an eagle chick, and go, well, one's amazing at what it does, and is perfectly adapted and beautiful and majestic, the other one is just this terrible fluffy thing that's bad at everything and needs to be protected."[4] However, if eaglets were really "bad at everything," then the entire eagle population would quickly disappear. A closer look reveals how their extra-fluffy feathers keep their small bodies warm, and the stretchy corners of their beaks help parents poke food in. Eagle parents are a huge part of an eaglet's environment, so it has adapted to make use of them. By contrast, many animal babies have had to adapt to environments that don't include any adults of their own species, much less their actual parents.

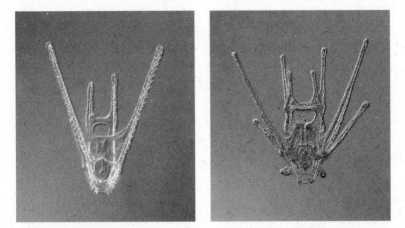

Thanks to the transparency of these four-armed (left) and eight-armed (right) pluteus larvae from a Caribbean sea urchin, their skeletal rods are visible—some simple and some fenestrated, or full of holes. The outside of their bodies are lined with cilia, which they use to swim and collect food.

Sea urchin larvae have adapted to look absolutely nothing like the adults that spawned them, and they are so wonderfully weird-looking that they were scaled up for use as aliens in *The X-Files*.[5] An adult sea urchin looks like a pincushion, with no front and no back, merely a top and a bottom. A larval sea urchin or *pluteus* (plural *plutei*) is shaped like a hand if it had an extra thumb to make it symmetrical. In fact, the similarity was so striking to me that while I was in Friday Harbor I designed the "hand pluteus," a drawing inspired by the classic childhood "hand turkey." You trace one hand, then place the other hand over it to trace an extra thumb. Since plutei are transparent, their internal anatomy is visible, so next you draw the larva's skeleton as little ladders inside each finger. Finally, you add its mouth and stomach, the latter digesting small green algal cells that were collected by the cilia lining each "finger."

The purpose of the pluteus's long fingerlike limbs presents a puzzle to science. Are they defensive spines, making larvae difficult to eat? Laboratory experiments in which plutei are readily consumed by small predators suggest not.[6] Do they serve as a

We don't all have access to a laboratory with beakers of eggs and sperm, but anyone with paper and pencil can create their very own larval sea urchin. Follow the steps from left to right and top to bottom to trace your hand, add the thumb of your other hand, and fill in anatomical details to produce a hand pluteus.

parachute to prevent the animal from sinking until it's ready? Or are they simply a way to increase the surface area covered with cilia, making it possible to collect more algae? Individual algal cells are quite possibly the most abundant planktonic food and the easiest to capture, each one tiny but their sheer numbers buoying up an oceanic ecosystem. Marine larval forms of many different phyla have evolved extensive arrays of cilia for algal capture, from the snail veligers we met in the last chapter to the flatworm larvae we're about to meet that look like demented plush toys.

In medicine, the most common flatworm is a tapeworm. In science class, it's little *Planaria*. However, the greatest diversity of flatworms is found in the sea: brightly colored and patterned adults with lobe-covered larval stages. Flatworm larvae are so weird and adorable that even when I was scrolling on social media, not even pretending to do research for this book, I came across a post with the caption "Can't get over what the larvae of some marine flatworms look like. Not totally sure you'll be able to handle it if I show you. If you think your [*sic*] ready you had better mean it" followed by half a dozen glorious images of the larval form known as a Müller's larva. These tiny swimmers are adorned with eight

Beneath its profusion of lobes, the Müller's larva of a flatworm is quite similar to a trochophore. Scientists suspect these similarities are a result of convergence—both types of larvae adapting to life as a tiny creature in a big ocean.

to ten large ciliated lobes, shaped like a rooster's wattles but much larger relative to the larva's body. Two lobes often hang below the others like round little legs, and many larvae have black eyespots that seem to gaze pensively at the viewer. Like plutei and veligers, Müller's larvae use their cilia both to swim and to collect algal cells. (The larva is named for Johannes Peter Müller, a nineteenth-century scientist who studied medicine and philosophy as well as fish, reptiles, and larval forms. He was a prolific author but does not seem to have been as poetically inclined as other embryologists. I could find no Müllerian verse.)

Cilia are so central to larval life in the ocean that we might be tempted to call them a defining characteristic, but being ciliated isn't enough to claim identity as a "true larva." Deciding what is and isn't a larva can be surprisingly controversial. Even Oyarzun's sibling-devouring bristle worm babies are not always considered deserving of the term. As she wryly points out, "In almost every paper we send to review, there's one reviewer who says, 'that's not a larva.'"[7] Scientific disagreement over the word "larva" is rooted in a difference of perspective between development and ecology. A larva can be a specific stage of development—a body shape distinct from that of an adult. A larva can also be a specific stage of relation with the world—a developing organism that interacts with an environment distinct from that of an adult.

For example, baby fish are typically called larvae, even when they look far more like a standard fish than a pluteus looks like a sea urchin. One reason for this is that larval fish can often occupy such different environments from adults that their parent-child relationship comes as a scientific surprise. Eel larvae found in the open ocean, far from the freshwater habitats of adult eels, were at first named a new animal: *leptocephalus*. When their true identity was resolved, the word *leptocephalus* was kept as a special name for the larval stage of eels (see insert, photo 10). In the history of larval biology this has happened over and over, as "species names" metamorphose into "larval stage names."

Even the word *caterpillar* predates the identification of these animals as larvae of moths and butterflies, as does the specific term *inchworm* for caterpillars of the Geometer moths. Moss animals, which are commonly found encrusting rocks and seaweed in tide

Adult horseshoe worms were described a decade after their babies (pictured here) had been named, and it took scientists another decade to connect the two as different parts of the same life cycle.

pools, were eventually discovered to produce a free-swimming ciliated larva—which had already been given its own Latin name, *Cyphonautes*, when it had been found earlier apart from adults. At least one researcher was so skeptical of cyphonautes as a larval form that he published a paper insisting it had to be its own animal.[8] Another group of larvae once had their own entire family, the Actinotrocha, to describe their unique appearance: like tiny hippopotamuses with tentacles. They are now known as the actinotroch larvae of tube-dwelling horseshoe worms (ah, yet another worm phylum).

Do humans have larvae? Although many biologists, myself included, habitually refer to our prepubescent children as larvae, it's a running joke rather than a biological statement. But human babies are certainly different from adults. Their habitat isn't entirely distinct, but while overlapping with adults in space they exhibit preferences that distinguish their behaviors and food sources. Adult humans tend to consume more bitter foods; children sweeter. One could argue that this allows human children to live in the same environment as adults without competing with them for resources. (Of course, baby humans also readily eat the food off their parents' plates.)

But even before they can run around and cram candy into their mouths, human infants are peculiar creatures. During the first year of life, a baby is more like an embryo. Compared to other great apes, all human births are premature. Based on our species' size and longevity, our pregnancies ought to last close to two years. However, if they did, the baby's head would grow too big to come through the adult human pelvis. So we evolved this deal where our newborns can emerge early and continue their basic physical development in the first few months out of the womb, sometimes referred to as "the fourth trimester."[9] And we evolved unfused skull bones to facilitate both our passage through the birth canal and our rapid postnatal brain growth.

Compared to other primates, this feature makes us uniquely human. However, compared to the rest of the animal kingdom, emerging into the wider world as a still-developing embryo is actually quite common. Many species of coral and urchin release their eggs into the sea to begin their lives as "orphan embryos," which are essentially premature hatchlings. They go through the embryonic stages from zygote to gastrula while bobbing on ocean currents, exposed to predators, chemicals, temperature changes, UV radiation, and more. Our human "extra-uterine embryos" are fortunate to exist in what is typically a much gentler environment. Coddled, carried, and catered to, human infants may seem at first glance as unimpressive as eagle chicks. But like the chicks, when we look closer, they display abundant adaptations: muscle reflexes, piercing cries, keen senses of smell and taste.

In lectures, Marshall says, "I used to show a photo of my one-year-old son, saying no one would think he's perfectly adapted to his environment, but he is." Human babies, like all animal babies, are geared for survival. "These aren't afterthoughts, they are the main game."[10]

The larval holobiont

Human babies, along with all other new life-forms, get busy becoming a mix of host and symbionts from the beginning. This makes each of us a *holobiont*, a term that encompasses the "complete" individual, host plus microbiome. Our earliest days are full of meeting new partners and establishing relationships, turning ourselves into an ecosystem. Humans pick up most of our microbial members by age three.[11] The most critical need for good microbes is in our gut, because gut microbes do far more than help us digest food. We're still figuring out the extent of their influence, but we know that the proper development of our immune system depends on chemical conversations with our gut microbes as well as microbes outside our body. In fact, the prevalence of immune-related health issues, like asthma and allergies and inflammatory bowel disease, in industrialized parts of the world has been linked to our increased germophobia and overuse of antiseptics. The "hygiene hypothesis" suggests that when we grow up in a microbe-minimal environment, our immune system can go haywire without enough partners to talk to.[12]

This is far from an argument against hygiene itself—I don't think any of us want to go back to the days before surgeons washed their hands. Even for us ordinary folk, an avalanche of evidence supports the health benefits of regular handwashing and avoiding contact with others when we or they are sick, and the COVID-19 pandemic certainly brought the utility of face masks for personal and public health sharply to the fore. The hygiene hypothesis is not about deliberately exposing ourselves to disease-causing pathogens or avoiding tools like vaccines and masks that help keep us safe. Rather, it's about connecting our bodies with beneficial microbes early in life. When I asked developmental biologists who had spawned how their work overlapped with their lives as parents, their responses zeroed in on microbes.

"Without my work I probably would have been much more of a clean freak with my children."[13]

"We tried to get our kids to go outside. I think that exposure to farm animals, exposure to dogs, exposure to dirt—we want our kids and our grandkids to get dirty."[14]

In this regard, we humans are representative of the animal kingdom. We are all holobionts, bearing microbes inherited from our parents or picked up from the environment or a combination of both. But what happens if you don't have a mouth to suckle with or shove dirt into?

Many invertebrates, like vent tubeworms, have nonfeeding larval forms. In the deep sea and near the poles, where sunlight is limited, algal food is also limited, and marine larvae are more likely to rely on yolk. These larvae often lack a mouth, gut, or both, until after metamorphosis. But this specialization isn't limited to extreme environments; quite a few shallow-water invertebrate babies live exclusively on yolk reserves provided by their mothers, possibly supplemented by the absorption through their skin of dissolved nutrients from the surrounding water. Scientists think that in most cases an actively feeding larva is the ancestral state, and nonfeeding larvae without mouths or guts lost them over evolutionary time. Sometimes species that are otherwise very similar diverge on whether they produce feeding or nonfeeding larvae. The reasons remain unknown, but they present an opportunity to study mouthless microbiomes.

Nonfeeding sea urchin larvae hatch from very large eggs as blobby babies that look nothing like the elegant multi-armed plutei of other urchin species. These gutless tots consume their yolk and quickly metamorphose into tiny juvenile urchins. The Australian biologist Maria Byrne wanted to know the implications of that. "The most important microbiome of animals is the microbiome that's associated with your gut. We've known that for years—it's important from humans to invertebrates. So what happens when you lose a gut?"[15]

She compared microbes in a pair of closely related Australian sea urchins, one with a feeding larva and one without. The feeding larvae, as expected, picked up bacteria from their environment, which then helped digest food and supply necessary nutrients to their host. The nonfeeding larvae were missing these microbes—not in itself a surprise, but what they had instead was a shock: *Wolbachia*, the same microbe that wreaks havoc with insect development.[16] What was it doing in the sea urchins? As it turned out, this particular strain of *Wolbachia* was deriving energy from the animal's yolk, and in return it was serving its host the same essential nutrients the regular gut microbes were giving to the feeding larvae. "We were flabbergasted," says Byrne. "It's opened my eyes. Animals don't live in the void of what's in their environment." Animal babies can coevolve different strategies with different microbes, based on what's available to work with.

The impact of bacteria and other microbes on development can hardly be overstated. With respect to insects, we find that our interaction with them is truly interaction with insect holobionts. The Japanese bean bug is a perfect example, as it picks up symbionts during its larval stage from the surrounding dirt. As its name implies, the bean bug is a pest of bean crops, and people spray insecticides to control it. Certain strains of soil bacteria can break down the insecticide chemicals, not only rendering them harmless but using them as a food source. Baby bean bugs that pick up these bacterial strains instantly become pesticide-resistant themselves.

Other insect larvae inherit symbionts from their parents, and this can get complicated. Pea aphids are a common agricultural pest on—want to guess? Of course, alfalfa! (Also, peas.) They can host a variety of different bacteria, which can in turn become infected with different viruses. However, the word "infected" creates the wrong impression—we think of infection as a bad thing. But when the pea aphid inherits the right bacteria, and the bacteria has the right virus, the virus releases a chemical that protects against parasitoid wasps. A successful wasp attack would spell doom for all three partners, so they work together to avoid it. (The viral chemical does its work, not by preventing the wasp from laying eggs in the aphid, but by disabling the development of any wasp babies that are laid inside the aphid. It's development all the way down.)[17]

"*Team* is the best metaphor I can find," says Gilbert, describing the need to think of each organism as a holobiont. "You can have the best goalie in the world, but if you don't have a good forward, you don't get into the playoffs. It's the team that succeeds, rather than the individual quarterback or goalie."[18]

Larval interactions with their own and other species

The microbes that larvae encounter can even determine their mating habits as adults. Scientists discovered that fruit flies prefer mates who ate the same food as they did when they were larvae, but this preference is driven completely by bacteria living within that food. If fly larvae are given antibiotic-treated food, as adults they exhibit no mating preference.[19]

This study, like the vast majority of fruit fly studies, was conducted in a laboratory. However, the entomologist Juliano Morimoto is curious about what fruit flies do in the wild. Fruit flies, after all, make their homes everywhere from jungles to kitchens. Morimoto wants to know how their larvae find food when it isn't handed to them on a petri dish, what they like and how they interact with each other. Despite their ubiquity in laboratory studies, Morimoto couldn't find any existing research on fruit fly larvae in the wild, so he set out to make some. "This is the most studied insect in the world ever in science," he says. "And we don't have this basic information about it."[20]

Morimoto's research has revealed that fruit fly larvae have not only feeding preferences but also social lives. We tend to think of social behaviors primarily in adult animals, who can compete for mates and cooperate to raise young, but interacting with peers has distinct benefits for larvae as well (see insert, photo 11). Gathering in dense groups on their food helps exclude microbes that would otherwise take over. A group of larvae also offers protection against predation, simply because any one individual is at a lower risk of being eaten. What's more, Morimoto told me to my great surprise, caterpillars of one South American moth can respond to predators by producing, as a group, an ultrasonic scream! It may serve as a warning or a repellent.[21]

Insects typically live most of their lives as larvae, and in some cases, they won't even eat as adults, so their entire life's nutrition depends on their childhood diet. Larvae take this charge seriously. Aquatic insect larvae can grow big enough to eat fish. Antlion larvae are so ferocious that the species is named for their habit of digging traps to catch and suck the juices out of other insects, especially ants. The adults, by contrast, are short-lived and forgettable. Even social insect larvae that spend all their time coddled inside a nest can provide for adults of the species. Some colonial wasps bring meals of meat to their larvae, who then produce a nectar-like saliva to sustain the adults—a fascinating reversal of our human tendency to prefer easily digested sugars in youth and more savory protein sources as we grow.[22]

Monarch butterflies and their relatives, meanwhile, depend on larvae not for nutrition but for protection. These butterflies are famously bad-tasting to predators, and the reason they're so nauseating is that as caterpillars they store up toxins from a diet of poisonous milkweed. Adults can't replenish this supply, as they feed on flower nectar instead. It was thought for a long time that each adult accumulated plenty of toxins during its own youth, but research in 2021 revealed that the adults of some species will attack caterpillars and suck their blood to supplement their own defensive store. (This behavior hasn't yet been observed in the iconic monarch itself, only in some of its close cousins.[23])

In the ocean, larval diet and adult diet tend to be more decoupled. Insect larvae are often larger or heavier than adults by the time of metamorphosis, then lose weight so they can fly. By contrast, marine invertebrate larvae are typically far smaller than adults all the way through metamorphosis. As we've seen, this is one reason these larvae often concentrate on collecting algal cells for dinner, while adults may be preying on whole animals.

Crown-of-thorns starfish, for example, are infamous for huge outbreaks of coral-eating adults that can destroy entire reefs. But crown-of-thorns larvae are tiny herbivores, collecting and eating algae and bacteria and even quietly absorbing nutrients dissolved in the water around them. Though this habit seems innocuous, it may actually contribute to outbreaks of adults. When agricultural runoff floods coastal waters with excess nutrients, crown-of-thorns larvae could grow and thrive and eventually settle in immense numbers.

Research on larval crown-of-thorns also offers encouraging possibilities for population control. These larvae contain toxins that were once thought to deter predators, because when scientists extracted the toxins and put them into fish food pellets, fish preferred to eat pellets without toxins. However, scientists were recently able to create a more ecologically relevant scenario, offering a variety of live starfish larvae to predatory fish. They found that many species of fish would readily consume large quantities of crown-of-thorns larvae.[24] Apparently, the toxins are insufficient deterrents when an entire delicious larva is on the menu.

Speaking of eating larvae, ribbon worms offer an interesting example of a group that contains both vegetarian and carnivorous babies, depending on the species. These wigglers live in or near water; you might spot them at a muddy shore or in a tide pool. Adults look like standard worms—until they find something to eat. Then they reveal themselves as terrifying predators by stabbing it with an enormous barbed proboscis. Meanwhile, baby ribbon worms look entirely unwormy. "There are some that look like hats, and there are some that look like socks," explains Svetlana Maslakova, the world expert on these strange animals.[25] Still other ribbon worm larvae look like no more than a "ciliated blob," yet they share their parents' capacity for abrupt carnivory. Maslakova and fellow worm researcher George von Dassow recently recorded the striking habits of one such larva. The adults of this particular ribbon worm species live on crab eggs (embryos show up as an important food source again!) and lay extremely tiny eggs of their own. When Maslakova found babies of this ribbon worm species in the plankton, they had grown to ten times the size of the eggs they had hatched from. She knew they had to be eating something, but what?

The answer came one day when some of her students attempted to extract DNA from the larval worms, and found crab DNA instead. It came from food being digested in the worms' stomachs, and it belonged to babies of the same type of crab that the ribbon worm parents preyed on. Maslakova was stunned, because these crab larvae, called *zoea*, seemed like impossible prey for tiny worm larvae to tackle. The zoea were much larger than the worms and covered with a hard armor-like exoskeleton. Finally, von Dassow observed the baby worms in action. Using a proboscis just like an adult, they struck the zoea and then climbed inside its shell, slurping it up from the inside like a living tongue. It's a tough life for baby crabs. If not devoured as embryos by adult ribbon worms, after hatching they face the ferocity of hunting worm babies.[26]

Larva-on-larva predator-prey action is found throughout the world, from sea to pond to leaf. One frog tadpole in northern Japan gets targeted by salamander tadpoles so often that it's evolved a flexible defense. When predatory salamander babies are present, frog tadpoles grow bulgy heads to make themselves harder to eat. It doesn't stop there, because salamander tadpoles in turn can grow wide mouths specifically to eat the bulgy babies. They only do it if frog tadpoles are around; otherwise, their mouths stay small, and they eat insects.[27] This ability to adjust body shape or behavior in response to the environment illuminates a valuable flexibility in animal development.

The bizarre behavioral and developmental adaptations of caterpillars

"It's not easy being a caterpillar," says the ecologist Martha Weiss of Georgetown University. "Lots of people want to eat you, and lots of people want to lay their eggs inside of your body."[28]

If you're thinking the latter part of that statement sounds like a reference to parasitoid wasps, then you are absolutely correct. After decades of studying caterpillars, Weiss says she isn't surprised if half the ones she collects in the wild turn out to be parasitized by wasps or flies. When they pupate, "instead of one butterfly, you get a thousand wasps." It's even possible for the parasitoids to parasitize each other inside the caterpillar. "It's a turducken of larval development happening at once," Weiss tells me, rather gleefully.

Although caterpillars are insects and technically have a hard exoskeleton, it's so thin and flexible it doesn't offer much protection from predators. However, some caterpillars have adapted it into an unusual defense. They have to molt as they grow, shedding the old skin (each molt marks a new *instar*, or phase of larval development). Many insect larvae hold on to their old molts in order to deter predators, and I agree it is quite a deterrent. I would not eat a mad hatterpillar, which stacks its old heads atop its current one (see insert, photo 13), no more than I would eat a golden tortoise beetle larva, which carries its feces as well as its old skins on a structure called an anal fork.

Other larvae cycle through a variety of predator-repelling tactics as they grow. The caterpillars of two swallowtail species at first resemble bird droppings, in mottled white and brown. After a couple of molts, they switch to mimicking snakes, vivid green with large imitation eyes. "You can't be bird poop if you're three inches long, but you can be a nice snake if you're three inches long," explains Weiss. These fake snakes even change their behavior, puffing up their shoulders and flicking out a pretend "snake tongue" to really sell the disguise.

Although Weiss is well versed in the world's diversity of caterpillars, she has devoted her career to one species in particular: the silver-spotted skipper. With characteristic dry humor, she says, "Nobody knows more about silver-spotted skippers than I do. Not many people know or care about silver-spotted skippers, so I'm not giving myself too great an accolade there."[29]

Weiss's work on this species has revealed fascinating adaptive characteristics. It is one of a few caterpillars that engineer shelters for itself from both the surrounding foliage and its own extruded silk. At each instar, a skipper caterpillar produces the same precise leaf shelter as other caterpillars in the same instar. They make the same cuts, the same folds, the same stitches with silk to hold the shelter's shape. They spend most of their time resting in this shelter, periodically emerging to eat and to expel their feces. And they don't just poop on the leaf or hang their butt over the edge and poop on the ground. No, they shoot their poop with incredible force over a distance that can be forty times their own length. This defensive maneuver avoids bringing their little house to the attention of predators who might see or smell an accumulation of waste nearby.[30]

After studying silver-spotted skippers for twenty years, Weiss is always ready to help new students learn to identify the five instars of caterpillar development. But she found these entomologists-in-training repeatedly coming to her with one particular problem: how to determine if certain caterpillars were in the third or fourth instar. "It always made me feel kind of incompetent when I would look at these, and I couldn't really tell what was going on," she says.

Finally, one of the scientists in the lab who was tracking the individual development of caterpillars day by day discovered the existence of an extra instar. Some of these caterpillars were going through a "third-and-a-half" instar, bigger than the third but smaller than the fourth. This extra instar only shows up in stressful conditions, when food or weather aren't optimal. But just because it's conditional doesn't mean it's uncommon—there are times when over half the caterpillars on a given plant will exhibit the extra growth stage.[31]

Weiss takes this experience as a crucial reminder that we limit our understanding when we hold on too tightly to what we think we know. Textbooks and published papers had all described five instars in silver-spotted skippers, so when Weiss and her students saw caterpillars in the extra instar, they kept "trying to mash them into one box or another." Many aspects of this caterpillar's biology are precise, rigid, and predictable. Each leaf-and-silk shelter is produced in narrow tolerances as though on an assembly line; each propulsive poo travels farther from the shelter the older and larger the larva becomes. These adaptations help protect skipper caterpillars from predators. The flexibility of their development is a different kind of adaptation, a baked-in capacity to cope with other kinds of environmental stress. It reminds us that adaptation isn't something that happened in the past and is now finished—adaptation can be flexible and ongoing.

If we looked only at all the adult animals on our planet and tried to imagine what their babies were like, we would probably settle on smaller, cuter versions of the grown-ups. I don't think any of us would come up with forms like plutei, Müller's larvae, or even caterpillars. Yet the very same sets of genes that produce the familiar shapes of adults also produce these tiny unfamiliar aliens—evidence of just how powerful, and how different, are the selective pressures of a baby's environment. Feeding and nonfeeding sea urchin babies show us that divergent forms and habits can evolve in the same environment with different microbes, and skipper caterpillars illustrate that individuals can adopt different strategies on the fly (or on the crawl) to cope with habitat variation.

Scientists may have described many details of any given species, but that doesn't mean we know everything it can do. And no matter how many features are consistent between individuals, each one can have a tremendous potential to alter its development during its own lifetime. When we allow them to do so, baby animals stretch and defy our expectations of the world, showing us that life is capable of far more than we typically give it credit for.

7

LESSONS FROM LARVAE

How Evolution Shaped Development and Vice Versa

They cling to youth perpetual, and rear a tadpole brood . . .
Live as tadpoles, breed as tadpoles, tadpoles all together!

—Walter Garstang, "The Axolotl and the Ammocoete"[1]

I have a confession: For years I struggled to understand evolution. Natural selection made sense. Genetics, too. I loved learning about inheritance, dominant and recessive genes, the fitness of different alleles in different environments. I was fascinated by stories of evolution in action, like the rise of antibiotic-resistant bacteria and soot-colored moths favored by the Industrial Revolution. But when I looked at the vast diversity of life-forms on the planet, I simply couldn't see how the mechanisms of evolution that I'd been taught could produce such major changes as extra legs, wings, and new senses.

It was hard to talk about my confusion because the topic of evolution was as rife then as it is now with political polarization. I didn't want to get into ideological or religious arguments. I just wanted to understand how natural selection could produce *macroevolution*—the large-scale changes from fish to lizard, hippo to whale. It was validating to learn, as I eventually did, that many other scientists had puzzled over this disconnect. The problem was most famously stated in 1904 by Hugo de Vries, the early Dutch botanist and geneticist: "Natural selection may explain the survival of the fittest, but it cannot explain the arrival of the fittest."[2]

The missing connection was finally made by advances in developmental biology, creating a whole new field called evolutionary development, or evo-devo. When I started graduate school, evo-devo was the hottest new topic in biology, and I eagerly ate up the research. The idea that changes during development could produce major novelties for natural selection to act upon wasn't new—Darwin's contemporary Thomas Huxley had talked about it—but evo-devo pulled back the curtain and showed us how it works. So far in this book, we've observed the many ways that evolution has shaped animal babies to suit their myriad environments. Now let's look through the other end of the telescope, so to speak, to see what these babies can teach us about evolution itself.

The intertwined history of embryology and evolution

Central to evolution is the concept of "common descent." Just like you and your cousins descend from a common grandmother, different organisms descend from common ancestors, further and further back in the family tree until all of us Earthling life-forms share the same common ancestor. This tree constitutes a *phylogeny*, a representation of evolutionary history. A phylogeny is built of relationships between groups, showing which organisms are theorized to be more closely related, having descended from a more recent common ancestor, and which diverged in the more distant past.

Embryology has always been a key informant of common descent. Darwin observed that the forelimbs of vertebrate embryos, whether arm or wing or flipper, are homologous—they have the same bones, suggesting that they derived from the same ancestral form. Therefore, all vertebrates are theorized to have a common ancestor. Darwin also observed that the barnacle, one of his favorite animals, produces a larval form called a *nauplius* that looks like a curious little cyclops. As adults, barnacles seem to have more in common with oysters or mussels, but their larvae link them to such crustaceans as crabs and lobsters, most of which go through a nauplius stage (though it is often contained in the egg rather than free-swimming). Thus, barnacles, crabs, and lobsters are theorized to share a common ancestor. And in 1866, a few years after the publication of *On the Origin of Species* (1859), the embryologist Alexander Kowalevsky found that the invertebrates known as sea squirts produce larvae that share many features with vertebrates. Adult sea squirts are headless, limbless, squishy baskets of flesh that sit on the seafloor filtering food particles out of the water, and they had previously been considered close relatives of snails. But free-swimming larval sea squirts have heads and tails and look like little tadpoles, allowing Kowalevsky to correctly reassign sea squirts to Chordata, the same phylum that contains us vertebrates.[3]

Embryonic forms were so useful for understanding evolution that in the same year, 1866, the biologist Ernst Haeckel formulated a statement that would inspire research and argument for over a hundred years. Using the word *ontogeny*, which can be considered a synonym for development, he claimed, "Ontogeny recapitulates phylogeny."[4] He meant that each animal in the course of its development passes through the stages of its species' evolutionary history. With a squint and some imagination, you can see that a human embryo superficially resembles a fish, then a reptile, and finally a mammal. This idea had already been around for decades, sometimes referred to as "the biogenetic law," but it was Haeckel's phrasing that stuck.

In the 1920s, the embryologist Walter Garstang took a new spin on the relationship between development and evolution. (Garstang put many of his scientific theories into poems, several of which I've quoted throughout the book.) "Ontogeny does not recapitulate Phylogeny: it creates it," Garstang wrote.[5] He was arguing that anatomical features of embryos, larvae, and juveniles can generate new adult forms, driving diversity on the phylogenetic tree. He coined the term *paedomorphosis* to describe species in which adults retain juvenile or larval features. This can happen either when sexual development accelerates so that juveniles start making eggs and sperm at a young age, or when sexual development proceeds at a normal pace but the juvenile form is never swapped out for an adult form.

The most iconic example of paedomorphosis is the axolotl: a newt that keeps its larval gills and aquatic lifestyle into adulthood. However, paedomorphosis is widespread. Some fly larvae develop ovaries and lay eggs while still crawling as maggots. These unfertilized eggs hatch into more larvae, which can continue the cycle, or they can go through metamorphosis to become sexually reproducing adult flies.[6] Even humans display paedomorphic traits, with our large heads reminiscent of the juveniles of other ape species. Garstang also invoked paedomorphosis to suggest that the entire vertebrate lineage might have evolved from the tadpole larvae of sea squirts.

Bizarre life cycles and parent-child conflicts come up frequently in one of my favorite sources of natural history humor, Underdone Comics. Rob Lang, the artist, kindly offered to illustrate the contrast between tunicate adult and offspring for this book.

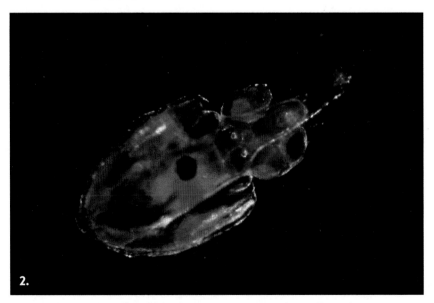

Baby fish, like this larval tuna, face high risks of both predation and starvation. They've evolved transparent bodies to hide from predators and gaping maws to tackle a wide range of prey.

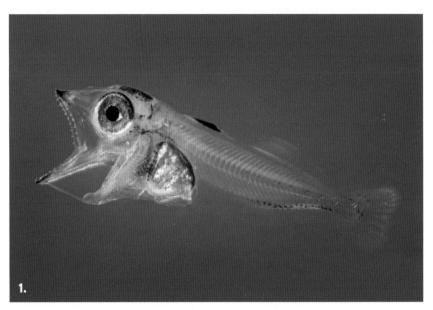

Born with only a handful of the color-changing organs that will one day cover its skin in the thousands, this baby Humboldt squid also bears a curious fusion of tentacles called a proboscis.

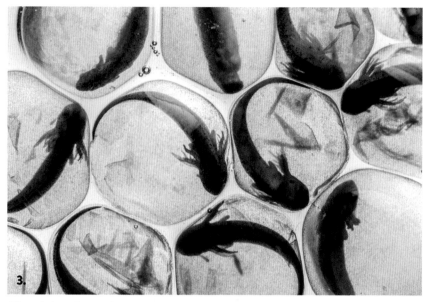

These baby salamanders depend on oxygen produced by algae that cohabit with them—not only inside their eggs but inside the very cells of their bodies.

1 mm

Deep-sea snailfish inject their eggs into various hosts that serve as safe havens for the development of the embryos. These babies were extracted from a xenophyophore, an enormous one-celled organism.

5.

The snail beside these blue egg capsules is not their parent—it produces differently shaped masses of bright green eggs. Such colorful pigments might serve as natural sunscreens, protecting embryos against UV radiation.

6.

These American robin chicks are altricial—naked, blind, and immobile—and yet incredibly well adapted to acquire the care they need, with specific calls to stimulate feeding and colorful insides of their mouths to provide a target.

A caterpillar of an Alcon blue butterfly smells like a baby ant, compelling ant workers to bring it back to their colony and feed it as they would one of their own babies.

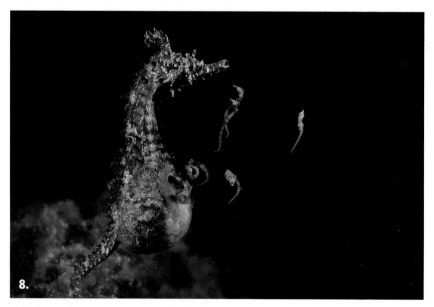

A father seahorse carries eggs in a sealed pouch and releases the babies after they've hatched inside him, making us question at what point "brooding" becomes "pregnancy."

9.

Many plant species produce seeds with lumps of edible fat attached—attracting ants, who bring the seeds back to their colony. The eggs of stick insects, pictured here, mimic these gift packages. Collection by ants disperses them to new locations and protects them underground until they hatch.

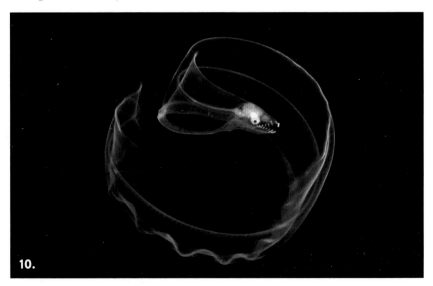

10.

Leptocephalus larvae of both American and European eels are born in the Sargasso Sea of the Atlantic Ocean. Somehow, through means still unknown, the growing babies know whether to travel east or west to reach the continent of their parents.

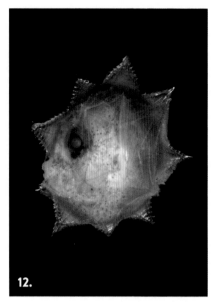

11. Caterpillars, often thought to be little more than eating machines, demonstrate a surprising array of social behaviors.

12. At first glance, an ocean sunfish larva looks like an alien spaceship. Closer inspection reveals its kinship with pufferfish and boxfish, a group of fish with scales modified into spines or plates.

13. The mad hatterpillar metamorphoses into a moth with an equally colorful name: the gum-leaf skeletoniser. It consumes the leaves of eucalyptus trees until only their "skeletons," or veins, remain.

Sunflower stars, listed as critically endangered in 2020, typically metamorphose from their larval form into juveniles with five arms. Unlike other sea stars, they add arms as they grow, as showcased by this ten-armed juvenile (a) and a twenty-armed adult (b).

A cicada nymph must anchor its feet firmly before beginning to struggle free of its exoskeleton (left). The black spots just below its red eyes (center) are concentrated pigment that will be gradually distributed throughout its body to produce the dark coloration of mature adults. A swifter process is the expansion of the cicada's wings (right), accomplished by pumping blood into the veins.

Bluehead wrasses begin their adult lives as small, yellow fish, either male or female. When a territory opens up, a yellow male or female (more often, a female) will transition permanently to a large blue-and-green male.

Comb jellies, which we met before as larval-gobbling mouths, also contain multiple cases of paedomorphosis. These beautiful animals look like transparent blobs with rows of cilia (the "combs") that reflect light in rainbow hues. They are the only adult animals that swim with cilia, despite the plethora of larvae that do so. There are three main groups of comb jellies, one of which, the cydippids, has two long, sticky tentacles. Only after the three groups of adults had been named did scientists discover that larval comb jellies of all groups have tentacles, too. This created a rare reversal of terminology, in which a larval form was named after the adult: cydippid larvae. The cydippid larvae of the two non-cydippid groups lose their tentacles as they grow, making it easy to imagine that today's cydippid adults evolved through paedomorphosis sometime in the past.

Meanwhile, it's not imagination but fact that larvae of several comb jelly species have been found producing functional eggs and sperm. Recently, a comb jelly in the Baltic Sea was discovered to have two populations: one that matures into reproductive adults, and another with reproductive larvae. The population made purely of larvae appears to be self-sustaining, requiring no input from the adults in the other population (the dream of many a feral kindergarten class). Scientists theorize that intense local predation drove the evolution of early maturity and reproduction in one part of the sea. When you're less likely to survive to adulthood, reproducing young has a greater advantage.[7]

How often might such a transition have happened throughout evolutionary history? Quite possibly more often than we realize. Turning larvae into adults could contribute to the "arrival of the fittest," but the classical embryology of Haeckel and Garstang still had no mechanism to explain how such changes might take place.

Then genetics came along, and embryology was cast out as not merely uninformative but unscientific. Thomas Hunt Morgan, the embryologist-turned-geneticist, wrote with the zeal of the converted, "Experimental embryology ran for a while after false gods that landed it finally in a maze of metaphysical subtleties."[8] Geneticists could finally elucidate the *how* of natural selection: Each gene had various states called alleles (like brown or blue for eye color), and the prevalence of alleles in a population could be observed, predicted, and calculated based on selective pressures and fitness advantages. This is how I was taught evolution, and this is where I got stuck. The frequency of brown and blue eyes in a population, while fascinating, does not inform the evolution of eyes.

In 1962, during the heyday of genetics, the well-known English biologist Sir Alister Hardy was writing the introduction to Garstang's posthumously published book of poetry. He pointed out that Garstang's great gift was recognizing the origins of evolutionary novelty in early development with his concept of paedomorphosis. "Siphonophores, ctenophores, cladocerans, copepods, insects and the very vertebrates, including Man himself, have all—and others too—been shown possibly, if never certainly, to have had a paedomorphic origin each from something very different. . . . I, for one, am confident that the coming generations of zoologists will judge it to be among the more fundamental conceptions given to our science in this century." However, he went on to say, "Of course none of Garstang's speculations can ever be established by experiment—they can never be proved as can an hypothesis in physiology."[9]

Now, in fact, they can.

Toolkit genes and an answer for macroevolution

In retrospect, I quite enjoy the fact that one of the groundbreaking discoveries that kicked off the science of evolutionary developmental biology was made in 1983—the year I was born. You might think, "Gee, Danna, you must have lived under a rock if you didn't learn about it until graduate school." And you're right! I did live under a metaphorical rock. But two other things also slowed my learning process: the fact that cutting-edge science takes a while to figure itself out, and the fact that cutting-edge science usually takes a long time to percolate into school curricula.

The key discovery, which was actually made in several labs in the early 1980s, was that simple mutations in a certain set of genes result in major changes to fruit fly anatomy. These mutations were present from the fertilized egg all through the childhood of the sugar-drinking maggot, but only manifested dramatically when the animal metamorphosed into an adult fly. Then, rather than determining something like eye color, these genes created new eyes where there had been none. They sprouted a leg in place of an antenna. They prevented wings from growing, or doubled the number of wings. Here at last was a significant source of evolutionary novelty for selection to act upon.

At first these genes, named Hox genes, were assumed to be limited to flies, but in subsequent years Hox genes were found in animals from frogs to humans, and it quickly became clear that they play a near-universal role in directing development. Although the fly mutants that gained the most fame were adults, Hox genes start working right at the beginning of embryogenesis. Small alterations to these genes can result in huge anatomical changes because each one plays many roles during development, and controls a wide array of other subservient genes. Sometimes called tool kit genes, Hox and other developmental control genes show how genetic "tinkering" can result in the kind of variation necessary for macroevolution.[10–14]

At the end of the twentieth century and the beginning of the twenty-first, the new science of evolutionary development—evo-devo—focused heavily on traditional model organisms. Discovery followed astonishing discovery. Analogous genes control heart development in both flies and mice. Eyes can be grown on one species by inserting eye tool kit genes from another. And despite the wild diversity of vertebrate body shapes, from giraffe to ger-bil, the same set of Hox genes divides each embryo into zones from head to tail and dictates what kind of anatomy should arise in each location—a head, a trunk, limbs. Any genetic mutation that affects where these genes go in a wee gastrulating blob of cells can shift the zones, with significant implications. Thus, for exam-ple, do snakes develop an incredibly extended trunk, no neck, and no limbs.

But what about the invention of new limbs, like wings? Very often this can happen by repurposing existing structures, as in the case of flying insects. Long before the discovery of tool kit genes and the advent of evo-devo, some people had theorized that insect wings came from ancestral gills. Numerous modern insects go through an aquatic stage—mayflies, for example, spend most of their lives as babies called nymphs in streams and lakes. During this underwater childhood, they need gills to breathe, and these gills look more like appendages than organs, protruding from the nymph's abdomen in serial pairs to extract oxygen from the sur-rounding water. The gills of other water-dwelling arthropods like crabs and lobsters are also elongated paired structures, although more hidden from view.

The theory that these gills could evolve into wings was eventu-ally proven by a beautiful confluence of work on organisms both model and not. Fruit fly work in the laboratory revealed tool kit genes for building wings, and the same genes were turned on in the gills of larval mayflies and even in the gills of crabs. When a gene is found across such distantly related organisms as fruit flies, mayflies, and crabs, it is most likely that the gene existed in the common ancestor of all these forms. Crabs now use the gene only to make gills, and fruit flies only to make wings, while mayflies have retained its dual use. Toolkit genes facilitate adaptation to different environments at different parts of the life cycle, also pro-viding a source of variation for future evolutionary developments.

Blueprints for bodybuilding

Moczek, the dung beetle specialist, remembers first learning that the process of development can teach us about evolution. "It was like this big door opened to a new universe," he says. "Not just that development is a product of evolution, but once in existence, it feeds back into where evolution is more or less likely to go. To me, that was mind-blowing."[15] Each stage of development, from embryo to larva to adult, is adapted to its environment, a product of natural selection. At the same time, every aspect of the developmental process, from gastrulation to turning on tool kit genes to metamorphosis, is a source of novelty that can be tweaked and tinkered with. As each organism builds its body from scratch, it has countless opportunities to build differently than bodies that came before. Random mutations and environmental modifications to genes could double the number of legs, stretch skin into wings, make eyes disappear, and express sensors for smells that haven't been invented yet.

What's more, development itself is one feedback loop after another. Cells, tissues, and organs come into being only in relation to each other. Growth is a group project, not independent assignments for each cell. The body contains ongoing conversations, cells influencing each other back and forth, producing a phenomenon called "correlated development." For example, our eyes develop when we're embryos because a bulge of brain cells meets a bulge of skin cells. The brain cells tell the skin cells to become the lens, and the skin cells tell the brain cells to become the retina. If they don't meet, neither becomes part of the eye.

You can see correlated development at work in different breeds of dogs—the results of artificial rather than natural selection. A dachshund, thank goodness, doesn't have merely short leg bones but also short leg muscles and tendons. A bulldog doesn't have only a short snout but also an appropriately reduced number of teeth and size of tongue. These intercommunications in the embryo can facilitate major evolutionary transitions. In 2014, a group of scientists showed that bichirs, a kind of fish with gills and lungs that can crawl out of water, develop a correlated set of physical adaptations when raised entirely on land. The terrestrial environment induces bone and muscle changes, as well as more effective walking behaviors. When considered in the context of evolutionary history, this suggests that ancient fish didn't need to figure out walking before moving onto land—the very act of moving onto land may have helped them to figure out walking.[16]

The integration of embryology and genetics allows researchers to tackle long-standing evolutionary puzzles, like the swallowtail butterfly's wings. Swallowtails are quintessential summer insects, often depicted in paintings and children's books, with long swoopy bits protruding from their hindwings. Scientists had been stymied by the question of how such a distinctive shape is grown during development. Moczek's mentor, Fred Nijhout, thought of asking whether the tail is really "grown" at all. He looked more closely and saw that during development the wings begin as much larger, rounded shapes.[17] Moczek summarizes the results: "Nothing is growing out there. It is programmed cell death, removing everything but the tail. It is like a cookie cutter." Programmed cell death is also responsible for the fact that a chick's feet are not webbed (but a duckling's are) and a human's hands are not webbed (but a bat's are). Early in development, we all start off with webs. A simple genetic switch determines whether the webs are kept or destroyed.[18]

Moczek has built on Nijhout's work with his dung beetles. Adult males of a few dung beetle species display distinctive and dramatic horns, and scientists had always assumed that the males grew them. That seemed almost too obvious to question. And because the presence of horns is so variable among dung beetle species, it had been thought to evolve multiple times.

Instead, Moczek and his colleagues discovered that every single dung beetle larva, male and female, of every species, produces a horn when it pupates. "And only very few of them actually carry that damn thing over to the adulthood, so *why bother*?" he asked. As it turns out, the horn is necessary for beetles to break free from their thick skins when they metamorphose, like the egg tooth a chick uses to escape its shell. The horn is a universal dung beetle trait, no doubt present in the group's last common ancestor. Most species evolved to resorb the horn after its purpose was served, but some encountered an adaptive value to keeping it, and it became repurposed as a flamboyant mate-attracting device for adult males—and in some species, adult females.[19]

Larvae: evolutionary relics or modern adaptations?

In the previous chapter, we addressed the question of why there are larvae from an ecological standpoint. Diverse, bizarre, and nearly alien larval forms exist because the young life stages of animals must adapt to environments that are different from those of adults. Their smaller size, sometimes drastically smaller, presents them with different physical challenges and dietary opportunities and makes them vulnerable to different predators. Now let us ponder the question of "why larvae?" from an evolutionary standpoint. Are the complex life cycles of so many animals today an evolutionary inheritance from a long-distant common ancestor, or have they emerged only recently and independently of one another? Across the animal kingdom, both scenarios can be found. Many larval forms arose as adaptations long after their adults had been around for a while—and also, some of the most well-known early animals had larvae.

Although far rarer than fossils of adults, fossilized larvae do exist, and the early developmental stages of trilobites are particularly well documented. Some trilobite species had a bulbous larval phase that likely swam in the plankton before settling to the seafloor.[20] But, wait, we can't put these larvae in an aquarium and watch them develop to adulthood, or sequence their DNA. How do we know they were trilobite larvae and to which adult species

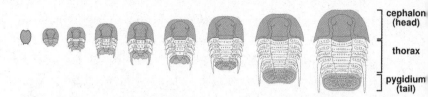

Trilobites began their lives as little heads, which soon added tails. As they grew, the midsection of their bodies expanded until they reached their adult shape and size.

they belonged? Because they were not the kind of larvae that look utterly different from adults; rather, they were a type of larva that scientists have nicknamed "head larvae."[21] They consisted, essentially, of a trilobite head and nothing more, which over time grew its trunk and limbs. This makes it relatively straightforward to match larva to adult.

Plenty of head larvae still swim in today's seas. The legs that a crab zoea uses for swimming will become its mouthparts when it grows up, as the entire larva becomes the head of the adult crab with the addition of a body and adult legs. Worm trochophores are also head larvae, their bodies lengthening as they grow segment by segment. Strathmann suggests that head larvae may have evolved as a way to start with the most important part of an animal, that part that can see and smell and eat, and make it immediately mobile. "Your head has to sprout wings," as he puts it. He imagines humans with such a larval stage. "In early life we would be mostly a head, maybe like those flying cherub heads in religious art. Did those Baroque artists have a special intuition about evolution of life histories?"[22]

Head larvae likely evolved multiple times in different groups, rather than existing today as echoes of the past. We know that larval forms are specialized to their habitats—plutei, veligers, and trochophores all develop bands of cilia, not necessarily because a small ciliated form was ancestral to all these animals but because it works so well for their planktonic habitat that natural selection favored it in each case. Another example is frog tadpoles and mosquito larvae.

They are offspring of, on the one hand, parents that hop on land, and on the other, parents that fly through the air. But both larval forms are aquatic, and they've adapted to their swimming lifestyle with similar long propulsive tails. They can be difficult to distinguish from one another when you're staring into a pond.

Nobody is proposing that mosquito larvae are vertebrates or evolved into them—when you look closely at their anatomy, it's obvious they lack anything like a backbone. However, the swimming tadpole larva of sea squirts shows more promise as a possible vertebrate ancestor, as it possesses a "dorsal nerve cord" very similar to our own spinal cord. Garstang sought to cast all of us bony beasts as a result of sea squirt paedomorphosis.

Another long-standing theory about the origin of vertebrates also made use of paedomorphosis, in this case from the larvae of lampreys. Lampreys are fish that lack jaws but are nevertheless very successful carnivores as adults. However, as larvae they look like simple worms, burrowing in the mud and filtering food from

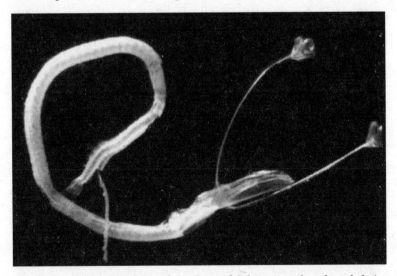

The dramatically stalked eyes of this dragonfish larva are thought to help it find food and avoid predators in the dimly lit regions of the sea where it lives (and tries not to die).

the water. These larvae, called ammocoetes, enjoyed the honorable status of "possible vertebrate ancestor by paedomorphosis" for many years. Then, in 2021, an incredible set of baby lamprey fossils revealed that the earliest lampreys didn't have ammocoete larvae. They looked like adults right from the start. The ammocoete, then, far from being a "primitive" form, is an adaptation of the modern lamprey life cycle.[23]

Complex larval forms and behaviors are some of the animal kingdom's most striking adaptations to life on Earth. We may think we're familiar with a species because we've seen the adults, only to find something entirely unexpected when we finally meet its babies. (For example, four different groups of normal-looking adult fish have independently evolved larvae with eyes on incredibly long stalks, sometimes nearly as long as the larva's whole body. As they mature, the eyes retract, and as adults you'd never know what they got up to in their youth; for another example, see insert, photo 12.) Figuring out how and why babies do what they do gives us an incredible window into the history of life on Earth. It's proving to be increasingly crucial to the future of life on Earth, as well.

RAISING THEM RIGHT

Conservation and Sustainability

We saw the risk we took in doing good,
But dared not spare to do the best we could . . .

—Robert Frost, "The Exposed Nest"

Humans have been raising the young of other animals since long before recorded history. We've raised them for food, for friends, and for fun. These days, we're raising a greater diversity of species, and for a greater diversity of reasons, than ever before. In large part, this diversification is motivated by a better understanding of our impacts on, and interdependence with, all the ecosystems of our planet.

Realizing that we've already driven many species to extinction, we design captive breeding programs to try to save the endangered ones. Finding that industrialized farming of mammals contributes to climate change and habitat destruction, we branch out into breeding new protein sources. With overfishing rampant throughout the world's oceans, we turn to aquaculture of our favorite species. Many of these efforts have had encouraging results, while others carry troubling implications along with their promise.

The early life stages of animals—eggs, embryos, larvae, and hatchlings—tend to be the most vulnerable to environmental impacts. Yet we rarely think about them until the problems are glaringly obvious: eagle eggs crushed because of DDT, salmon fry struggling to survive in contaminated streams, sea turtle hatchlings disoriented by light pollution. But once we recognize the situation, we can often help. Science and legislation brought back the eagles, various conservation efforts are underway for salmon, and turtle-safe lighting along with other nest protections are helping some sea turtle populations to rebound. One of my favorite stories of conservation success is the California condor.

"We will have to cuddle with this egg"

When I was born, in the early 1980s, there were only twenty-seven California condors in the world. One of my most vivid memories of visiting the Los Angeles Zoo as a young child was climbing up on hot metal bleachers to watch a bird program in which staff told us about efforts to rebuild the condor population. At the time, I was probably shorter than an adult condor.

Condors are enormous birds, their wings stretching 10 feet (3 m) and their weight about that of a human toddler. Of course, they're born smaller than that, but the chicks grow to nearly adult size while still dependent on their parents for food. I was astonished when watching a "condor cam" video to see a condor parent swoop into a cliffside nest site to visit a chick that looked, to my untrained eye, every bit as big as the parent. (There's no obvious visual difference between males and females, so I couldn't tell if it was the mother or the father.) The chick's head, however, was still covered with black feathers, in contrast to the red-skinned head of its parent. When the parent approached and opened its mouth, the chick promptly jammed its entire feathered head inside. The parent maintained a steady upright stance with mantled wings, as the baby frantically flapped and jerked to get its dinner.

I called the wildlife biologist Joe Burnett, who's been working with condors for over twenty years, to talk about what I saw on the cam and the overall importance of chicks in the conservation of this species. Human impacts on the early life of condors have been severe, from DDT thinning their eggshells to plastic trash filling their guts. Plastic bottle caps and other bits of garbage aren't eaten accidentally—parents collect them on purpose for their chicks, because they look like chips of bone or seashell. These natural sources of calcium are like vitamins for baby condors. "There was a chick in Southern California that was just full of bottle caps, and had to be euthanized," says Burnett. But the main threat facing condors is lead poisoning. Many of the carcasses that they scavenge were shot by human hunters, and the ammunition left behind can be fatal. "We're always devastated when we have a pair raising a chick and one of the pair dies of lead poisoning. We had that this year, in two nests. When you only have six nests, that's kind of a big deal."[1]

Although losing a parent in two out of six nests is crushing news, even the fact that condors are nesting in the wild again seems like a minor miracle. In the 1980s, the species' entire population of twenty-seven was captured in the hopes of rebuilding greater numbers in captivity. It worked. Over time, condors hatched and raised in zoos were released to create several stable populations in the wild, and the total count as of 2020 was 504 individuals.[2]

Burnett has been an integral part of this program, coordinating the first condor releases in central California in the 1990s and starting a captive breeding program at the Oregon Zoo in 2003. When he first took on the responsibility of incubating eggs in captivity, he says, he acquired a new respect for condor parents doing it in the wild.

Burnett recalls the media attention lavished on the first condor egg laid in captivity at the Oregon Zoo. "*The Oregonian* followed this egg like it was some rock star. I was getting calls every day from the lead reporter." As is routine in captive breeding, he removed the egg from the pair that laid it, encouraging them to lay and incubate a second egg while the first was artificially incubated. This technique doubles a pair's potential reproductive output. Burnett had set up a brand-new facility to receive the egg and was living on-site with his wife.

Then an early season ice storm shut down the city of Portland. "So power goes out, and I have an egg in an incubator, and we have to keep it going. I go out to pull the generator, I pull the cord and the cord breaks." Burnett's wife suggested heating water on their backpacking stove in order to keep the egg warm with a hot water bottle, so that's what they did, continuously heating more water as the bottle cooled, monitoring the temperature with an analog thermometer. When they began running out of fuel, Burnett says, "I told my wife, this egg, if we can't keep it warm, our body temperature is 98.6, I said, we will have to cuddle with this egg. And she was like, let's do it."

Just before it got to that point, news of the condor egg's plight reached Portland Gas and Electric. They immediately prioritized restoring power to the incubator, and the egg was once again well situated. With the crisis of incubation over, the next question was "Who would care for the baby after hatching?" The original parents had indeed produced a new egg and were busy with that. The only condor pair available to hatch and care for the ice-storm egg had never laid an egg of their own and had no experience with babies. Burnett gave them the egg, hoping for the best, and the adoptive couple took to parenting like a condor to thermals. They successfully hatched and raised the chick, and he was released into the wild in 2006 at Pinnacles National Park in California. Burnett says, "He is now the top dog. He's raised six chicks now; he's breaking the curve in terms of survival. Every time I see him out in the field, we have a special bond." Burnett amends this with a laugh. "He has no idea."

As you can tell, raising a single chick is a huge investment for condor parents. They reproduce more like large mammals than like other birds. This effort is what allows the chick to grow from the size of an avocado to the size of an adult in six months. In addition to continuous feeding, the parents cuddle and play with their young chicks. Burnett admires the joint investment of both parents. "It's really beautiful to watch these pairs raise chicks because it's really equal. If more humans could be like that."

After the first couple of months, the cuddling tapers off. The parents need to spend more time foraging for food, and the chick needs to be encouraged to get out of the nest, where it's a sitting duck, so to speak, for predatory ravens, owls, eagles, and even other condors. In this context, Burnett explains the behavior of the older chick I saw on the condor cam.

As more captive-born California condors are released into wild breeding populations, more chicks hatch in nests on cliff faces or redwood trees. However, zoo breeding remains integral to the species' conservation and recovery.

"They're left alone a lot. When the parents show up, they're incredibly excited, like someone that's been in captivity or on an island. It's the only social interaction they get, and they get fed. It's almost violent, the way they feed. It's super primal." (Honestly, it reminds me of my newborn human latching on to my nipple on her first day, and a nurse in the hospital commenting, "I see you've got a little barracuda there." Oh, the ferocity with which tiny creatures seek sustenance!)

Another change in condor parenting comes when the chick begins learning to fly. Now parents become guides and teachers. Burnett has watched condors on the central California coast nesting in both redwood trees and on cliffs, and he observes differences in learning to fly between these two environments. The cliff sites are more open, which usually lets the chicks learn faster, while redwoods slow them down. Each location has its advantages. A fast entry to the world of flight means more crashes, like a baby accumulating bruises in its hurry to walk—but walk it does, and quickly. A slower approach offers more time to build both muscles and confidence, though it can take longer to reach competence.

Because one of the risks to the still-precarious condor population is West Nile virus, field workers carry out a program of vaccination. All condor parents have now been vaccinated, and mothers pass some antibodies into their eggs—another investment in her offspring, along with yolk. But these antibodies don't last long, so biologists aim to vaccinate the chicks as soon as they can reach them. They face the challenge of not merely accessing remote nests but also dealing with the chicks' disgustingly effective tactic for deterring predators. "When you approach, they regurgitate whatever they've been fed by Mom and Dad," says Burnett. "It's pretty gross. It's not only rotten meat, but it's partially digested rotten meat, and it's putrid."

On one vaccination visit to a chick in Pinnacles, which is about 40 miles (64 km) inland, Burnett whiffed a familiar flavor of putrid that he hadn't expected: marine mammal meat. This chick, who had never seen the sea, had regurgitated a meal of decaying whale or dolphin that a parent must have scavenged from the distant coast. *How cool is that?*, thought Burnett. *These inland birds have a connection to the ocean.*[3]

Mayflies: a lifetime of youth and an eyeblink of adulthood

Condors spend nearly a year as chicks, fully dependent on their parents, and take several more years to reach maturity. However, given that they can live up to sixty years or more, that's a relatively small fraction of their lives spent as babies. Even so, it is the most precarious and important part of their life cycle in terms of survival.

Young mayflies, or nymphs, breathe water through the delicate gills that look like feathers on their abdomens. Three elegant tails may make them look larger and less tempting to predators.

By contrast, many other species spend nearly all their lives as babies. Remember that some insects don't even eat as adults? One of these is the mayfly, popularly thought to live for only one day. But each adult that seems to live and die in twenty-four hours has already lived for a year or more as a larva on the stream bed. (In other words, the childhood of a mayfly and the childhood of a condor are comparable in length!) We met larval mayflies or nymphs in the last chapter because of their gills. They are as aquatic as fish—and are often eaten by them. The fortunate few who survive lay hundreds to thousands of eggs, of which most will once again succumb to piscine predation.

The numbers involved might make survival sound like a matter of pure chance, and for many years people did think of larval life as a lottery. A crop of babies is thrown into the environment to either flourish or perish on luck alone. However, whenever we look closely, we find ways that larvae can and do influence their own survival. In fact, my first independent research project as an undergraduate focused on the behavior of mayfly larvae. Despite the obvious field work potential, I completed the study entirely indoors on a computer.

It started with an ecological modeling class. This did not teach me to strut down a runway dressed in leaves and seaweed (haute ecouture?) but to use mathematical equations to predict and test real-world interactions. I got a summer job in my professor's lab, creating a computer model of mayfly larvae in a stream. It shared some features with PLD models of larval dispersal, but the one-way flow of the stream was significantly easier to program than complex ocean currents. In addition, mayfly larvae aren't plank-tonic—they cling to the streambed, grazing on algae. However, the water does carry some individuals away, giving rise to the "drift paradox": How can a larval population persist in the face of constant loss downstream? With the help of a patient mentor, I showed that larval behavior can solve the paradox—at least in a computer model. We knew from field studies that mayfly larvae were more likely to enter the current when there was less algae available on the streambed and more likely to stay put when there was plenty to eat. When I incorporated this algae-dependent behavioral drifting into the model, mayfly populations thrived wherever there was sufficient food. Our answer to "How can a larval population persist?" was "Sometimes it doesn't," but not due to chance. If larvae have enough control over their entry into the current, they remain in areas of good eating and disperse away from poor fare.

Baby mayflies may not be as obvious subjects for juvenile literature as baby bears or ducklings, but I was so enamored of my research that I felt compelled to write a short story in the style of a children's book. It begins: "Effie was born Upstream. She was the two thousand three hundred and ninth of two thousand three hundred and seventeen larvae that hatched from the eggs laid by one mother mayfly. Effie had one thousand two hundred and eighty-three older sisters and one thousand and twenty-five older brothers." Crowded out of their Upstream location by too many larvae competing for not enough algae, Effie and her two favorite siblings decide to "drift jump" farther Downstream to look for better pastures. Despite a close encounter with a hungry trout, all three survive to the end of the story, but only because they're fictional protagonists. I knew the grim reality facing the majority of larval mayflies.

We might be inclined to discount the vast number of mayflies who die in their youth as irrelevant. After all, they're not contributing to the next generation. But they are crucial to the ecosystem; their abundance supports healthy populations of trout, dragonflies, and crayfish. Young mayflies, like young humans, are also more sensitive than adults to environmental toxins. (Although in the case of mayflies, it may be simply because adults don't have enough time to express sensitivity in their incredibly brief maturity.) For decades, people have used mayfly nymphs as indicators of pollution in streams and lakes.[4] Western Lake Erie, for example, was once home to 30,000 tons of mayfly nymphs—nearly a trillion individuals. Then, in the postwar boom of the mid-twentieth century, excess fertilizer in agricultural runoff fed explosions of planktonic algae. Nymphs can't eat these algae, and the blooms were digested by microbes that used up all the oxygen in the water, killing off nymphs until none were left by 1960. Scientists recognized the loss of mayflies as a sign of a damaged ecosystem, and conservation measures to reduce nutrient pollution were put in place. By the late 1990s, the mayflies were back up to their previous numbers, reassuring managers that the ecosystem was back in balance. However, mayfly recovery might owe as much to the efforts of humans as to the impacts of an otherwise unwelcome species: zebra mussels, which proliferated throughout the Great Lakes around the same time that people were trying to reduce pollution. This invasive species is very effective at filtering algae from the water and transferring nutrients from the plankton to the bottom, where mayfly nymphs live. The potential positive impact of zebra mussels on mayflies reminds us that no species is morally "good" or "bad"; all are players on the ecological field with complex and cascading interactions, always subject to change as conditions shift. Unfortunately, after reestablishing themselves thanks to both humans and zebra mussels, some mayfly populations are now declining again due to the effects of climate change.[5, 6]

The mayflies that do make it to maturity contribute not only to the next generation of mayflies but also to the next generation of songbirds. Although their adulthood is transient, their abundance is staggering. Within the span of a few hours, literal billions of insects can fly up from large waterways like Lake Erie or the Mississippi River. Tree swallows and phoebes build their nests nearby, and they time the hatching of their eggs so they can feed mayflies to their chicks. Thus, while adult songbirds are hardly aquatic animals, their babies rely on healthy underwater habitats, where mayfly nymphs can thrive in sufficient numbers to emerge as a flying feast at the end of their lives. The loss of one species' babies affects other species' babies, rippling through the food web.

Poop and plastic: how insect larvae consume our waste

It's interesting how different the babies of different insects can be. While mayfly larvae are sensitive youth that require clean water and specific algae, black soldier fly larvae are tough janitors that can be grown almost anywhere and eat almost anything. Like other fly larvae, they are a kind of maggot, resembling thick and bristly segmented worms, white when young and darkening to yellowish brown as they grow. Entomologist Jeff Tomberlin compares a black soldier fly larva to the Hulk. "Whatever you throw at it, it powers through." He encountered a remarkable example of their speed and voracity when one of his students had some larvae in an outdoor enclosure, which kept tempting birds to feast on his experiment. One day Tomberlin came across a dead bird. "I thought, I'm going to play a joke on my student. So I picked the bird up and put it in the bin with the larvae. And I said, 'There's a bird eating your larvae!' And he ran out and said, 'Where?' And I ran out, and it was gone.

Within thirty minutes, the larvae had come up, taken the bird carcass down, and had eaten it. It was a little sparrow. They just ate it."[7]

The result of such insatiable munching is that a black soldier fly larva will grow to fifteen thousand times its size between hatching and metamorphosis. That's the equivalent of a 6-pound (3 kg) human newborn growing to the size of a 90,000-pound (40,800 kg) fin whale. In the wild, black soldier fly larvae complete this staggering growth in about ten days. Optimized conditions in the laboratory can get that down to seven.

Tomberlin describes his "baptism" into the world of black soldier fly research at an industrial poultry house in the 1990s. About a hundred thousand chickens were housed in the enormous building, elevated above a basement area where their droppings piled up. Black soldier flies thrive in this environment, making it an excellent place to collect them for laboratory research. Tomberlin had gone with his adviser, and he says, "I remember looking in there, and it's just raining manure, and I went, 'Dr. Shepherd, we have to go in there?' And he said, 'No, Jeff, *you* have to go in there.'"

I ask if Tomberlin managed to get the larvae. "Five-gallon buckets of them," he confirms.

Today, black soldier flies support a brisk business. When I went online to order some for my compost pile, I had a dozen different companies to choose from, some specializing in black soldier fly larvae as pet food, some as farm feed, others as biowaste treatment. The industry owes its existence to

Black soldier fly larvae are the opposite of picky eaters. Grown on diets of compost, manure, or corpses, they themselves can then be fed to farm animals—or, hypothetically, to humans.

the rearing techniques developed by Tomberlin and his colleagues in their laboratories, which were conducive to scaling and continue to be refined. Tomberlin is proud of what's come from his work and sees insect larvae as a solution to a variety of ills, from greenhouse gas emissions to food insecurity.[8]

I was baffled by one particular application of black solider fly larvae: the decomposition of manure. Manure is just animal poop, which the larvae eat and then produce . . . more animal poop. This secondary poop, called frass, is touted as an excellent fertilizer, but where's the benefit over simply fertilizing with the original manure? Tomberlin explains that raw manure from birds or mammals has several serious drawbacks. It often contains pathogens, like *E. coli* and *Salmonella*, which can then contaminate any food grown in the area. It's also very wet, which makes it heavy and costly to transport. A facility like the poultry house Tomberlin visited is most likely creating more manure than nearby farms need as fertilizer, but it's difficult to ship elsewhere. When black soldier fly larvae digest it, they typically kill off pathogens and take out moisture, simultaneously making the product safer and easier to use.

These larvae can perform a similar service in compost piles, whether industrial or small-scale. Tomberlin compares a compost bin to a furnace. "You're stoking it with food waste and the larvae are digesting it. They're the fire." The larvae even reduce the amount of greenhouse gases emitted in the composting process by adjusting the microbial composition of their environment. They eat or outcompete bacteria that produce carbon dioxide. Their impact on microbes can extend to clearing out bacteria that are carrying antibiotic-resistant genes, making it less likely that such genes will be transferred to pathogens and end up causing human health problems.

The digestive superpowers of larvae are superpowers of the holobiont, a result of cooperation between host and microbes. Recent research suggests that the right bacteria-larva combination can even break down plastic. Despite its rapacious habits, the black soldier larva doesn't play host to plastic-munching bacteria. Instead, it's the humble mealworm.

Mealworms are baby beetles, which are also produced and sold commercially as farm and pet food. They're a bit longer and rounder than fly larvae, with dark stripes between their segments. I remember buying mealworms to feed my pet turtle as a kid, and my own children are now fascinated to watch our neighbor feed mealworms to the birds and lizards in her yard. But despite their utility, mealworms acquired their name in the 1700s as pests, eating through the "meal" or grains that humans had stored up for their own consumption.

The phenomenon of plastic-eating mealworms was also initially seen as a pest problem. In the 1950s, plastics were considered wonderful new materials, and larvae that could chew holes in plastic bags were a real problem in food packaging. People began to study them with the goal of designing pest-resistant plastic. At that time, no one was interested in figuring out whether the mealworms were actually digesting the plastic, explains the materials science researcher Wei-Min Wu. Just because an animal eats plastic doesn't mean that it breaks down in the gut, as we saw in the heartbreaking case of the plastic-filled condor chick. "Lots of animals eat plastic," points out Wu. "My mother raised chickens; chickens eat plastic. Birds that eat plastic die."[9]

Mealworms, however, did not appear to be made ill by their consumption of synthetic polymers. The first evidence that these "pests" could not only damage but digest plastic came from a teenager's science fair project in China in the early 2000s. Over the years, more science fair projects around the world returned to the question, from 2003 research on mealworms eating Styrofoam to a group in 2009 that managed to isolate the responsible bacteria. (It remains a popular topic for children and teens to tackle today.) In 2015, published scientific research confirmed mealworms' ability to break down Styrofoam. Since then, Wu and his colleagues have showed how mealworms and their gut bacteria break down the complex polymers of Styrofoam as well as other common plastics, excreting not smaller bits of plastic, but plain organic waste.[10]

Why can mealworms do this? It's not like they've evolved alongside plastic for millions of years. However, unlike black soldier fly larvae, which eat mostly animals and animal products, mealworms eat plant matter, including the difficult-to-digest plant protein lignin. Lignin is a major component of wood and can't be broken down by human guts or, indeed, most animal guts. Insect-bacteria symbioses that evolved to tackle lignin may be easily adapted to plastic. Wax worms, a kind of baby moth that eats wax in the wild, are another example of larvae that may be preadapted to digesting plastic.

But right now mealworms are the more promising area of study, because these insects, called darkling beetles when they grow up, are incredibly abundant and diverse. ("Remember the movie *The Mummy*, the Egyptian in the tomb? That's the darkling beetle," says Wu.) There are at least two thousand species of darkling beetle, and most of their babies haven't been studied yet. Each probably contains its own characteristic bacteria, and that's a lot of variation to work with—both in the laboratory, as we humans seek solutions to plastic pollution, and in the wild, as natural selection favors those organisms that adapt to their human-modified environment.

We can't yet send a larval cleanup crew out into the world to handle all our plastic waste. Wu doesn't see that as a likely eventuality, either, given the complexity of the larval-bacterial system and the difficulties of scale. But chemists could use knowledge of how insect guts break down plastics to modify plastic products and create more effectively biodegradable materials.[11] Meanwhile, industries keep churning out mealworms, wax worms, and black soldier fly larvae for yet another purpose: dinner.

Farming fussy babies in the sea

Insect protein is sometimes touted as a solution to feeding Earth's growing human population. When it comes to eating insects, it's usually larvae that are the best source of nutrition (as caterpillar-eating birds would agree). In metamorphosis, they'll turn a lot of good fat and protein into far less nutritious wings. Of course, this isn't a "new" food at all; humans have long been eating insect larvae. But as the wild fly researcher Morimoto points out, "If you go to an indigenous community and they tell you they eat larvae, the first thing you're going to do as a colonizer is try to get rid of that behavior." In his passionate paper, "Why Should Humanity Care About Insect Larvae?" Morimoto makes a cogent case for farming insects based on cost effectiveness and nutritional value.[12]

The obvious need for practical approaches to worldwide nutrition has led to a renewed focus on invertebrate consumption across ecosystems. Insects are far from the only invertebrates that people can eat. In fact, many humans who are squeamish about eating terrestrial invertebrates have no qualms about consuming aquatic ones like shrimp, lobsters, oysters, and calamari. But are these protein sources as sustainable as insect farming is, or as it is claimed to be?

When it comes to overharvesting from the sea, we've dealt our worst damage to vertebrates, from whales and turtles to sharks and salmon. By contrast, many invertebrate species seem practically invulnerable to overfishing. They produce scads of babies, are flexible feeders, and have a quick generational turnover. Unfortunately, unchecked human interests have indeed managed to devastate numerous marine invertebrates. White abalone, for example, were once caught and eaten in the hundreds of thousands of pounds. But a few years of that cleaned out the stock. In 1997, the state of California closed the fishery completely for fear of extinction, but white abalone remain endangered in 2022 because of their reproductive habits. Adults are broadcast spawners, which means they release their eggs and sperm freely into the water to mingle. Fertilization depends on proximity to other members of their species. The few adults that remain are too spread out to reliably produce babies. So, to restore populations of white abalone, people have been raising them in captivity, just like condors. And just like condors, the earliest stages have been the trickiest to keep going. Larvae and juveniles, as it turns out, have very specific needs, especially when it comes to temperature.[13]

Raising abalone in the laboratory can have the dual benefit of restoring wild stocks and producing meat for human fare. Aquaculture, which means farming seafood rather than fishing for it, sure seems like a more sustainable option. But many highly prized species are also highly challenging to raise. Flying squid, including my beloved Humboldt squid, are widely eaten marine invertebrates. Being able to farm them in captivity, rather than having to capture them in the wild, would be a tremendous boon to calamari lovers around the world. But I found their embryos' sensitivity to temperature derailed my efforts to rear them. Humboldt squid breed in tropical and subtropical water, so they must have adaptations to cope with the heat, but the warmer the water I tried to grow the babies in, the bigger my issues with fungal infection and egg mortality.

Rachel Collin, who studied cloning larvae in the Caribbean, has been frustrated by the challenge of raising warm-water babies in her laboratory. She notes with some humor that many of us embryo and larval biologists were trained at Friday Harbor, where the water typically ranges from 42 to 55°F (6–13°C)—a far cry from the 77 to 85°F (25–30°C) water in the tropics. "At Friday Harbor, you change the water every other day. Maybe [for tropical water] what you need to do is change it three times a day, but who's going to do that? You stick with changing it every other day, like everyone always does."[14] Raising embryos and larvae in warm water can lead not only to infections of the animals themselves but also to problems with their algal food. Some types of algae produce toxins at high temperatures. Some just die.

I had an even more basic problem with feeding my Humboldt hatchlings. I offered them brine shrimp, algae, and salads of local plankton. Knowing that many invertebrate larvae can absorb dissolved organic material from seawater, I added new water frequently, hoping they might gain some nutrition thereby. Alas, I was never able to successfully feed the babies from the Humboldt egg mass, nor any that hatched from my forays into artificial fertilization. Meanwhile, plenty of other squid species will eat in captivity right after hatching—including the opalescent squid, which we also studied in our lab. These babies are as predatory as their parents, using their eight arms and two tentacles to attack live copepods and shrimp. This brings up a sustainability issue with raising any kind of squid for human consumption: unlike black soldier fly larvae that eat compost and manure, squid must be fed substantial quantities of live prey, typically shrimp and fish that would make perfectly good food for humans. Converting all that edible matter into squid flesh before we consume it is inefficient, at best. Many people argue that we should take a step even further down the food chain, past those shrimp and fish that would need to be fed something, too. And abalone aren't predators, but they eat masses of algae. Why not derive our own protein directly from algae? As a vegetarian myself, I confess that's what I'm most interested in.

But that didn't make me any less excited when, in 2018, a group of scientists, including the mentor who had taught me *in vitro* squid fertilization, finally figured out the diet of Humboldt squid paralarvae. They did not manage successful laboratory feeding (to the best of my knowledge, that still has not been accomplished in 2022), but they collected and carefully analyzed paralarvae from the wild. Now, the paralarvae of Humboldt and other flying squid are anatomically unusual among squid paralarvae. Adult squid have eight arms and two long elastic tentacles that can shoot out in a high-speed "tentacular strike" to capture prey and bring it back to be restrained by the arms as the animal consumes it. Opalescent squid

PART III

COMING
OF AGE

METAMORPHOSIS
But Happier Than Kafka

Although it is night, I sit in the bathroom, waiting.
Sweat prickles behind my knees, the baby-breasts are alert.

—Rita Dove, "Adolescence-II"[1]

Although puberty can feel like a catastrophe, humans don't undergo the type of "catastrophic metamorphosis" that so many other species experience. We don't pupate or spin a cocoon. We don't grow adult bodies that burst out of our abdomens, nor do we turn ourselves inside out and eat our childhood selves. And yet Scott Gilbert, author of multiple developmental biology textbooks, says, "I think humans undergo metamorphosis—less dramatic, but it's still pretty dramatic." He lists the similarities between adolescence and metamorphosis: changes in body shape, muscles, gonads.[2] Perhaps most important is the significant shift in environment that we experience.

"What is metamorphosis?" asks Gilbert. "It's going from one environment to another environment. And that's what adolescence is. Our environments are social as much as they are natural." We've all experienced pubescent upheaval in our bodies, and we probably all remember how the physical transformation of puberty felt deeply entangled with our shifting social-emotional world.

Consider the far more extreme anatomical changes faced by so many other young animals, and the environmental shifts that accompany them. A caterpillar dissolves into goo, only to reform as a butterfly—a shift from land to air. A ribbon worm larva grows an adult inside itself, which eventually eats the larva—a shift from ocean currents to rocks and mud. Deep-sea worm larvae that land on sunken whalebones become females, while those that land on female worms become males—a shift from sunlit surface water to lightless depths.

How do such seemingly magical transformations occur? And, crucially, how do animals and their environments interact to produce metamorphosis at the right time and place? These changes occur over a very brief portion of an animal's lifetime, but they are watershed moments. Each metamorphosis forges a living link between one environment and another.

Catastrophic change

The term *metamorphosis*, like *larva*, has a range of meanings. It can cover minor changes in crustacean appendages that facilitate a switch from swimming to walking. It can refer to pluteus harboring a rudimentary urchin that pops out and destroys the larval body. It can even happen more than once in a life cycle—parasites with multiple hosts often go through a distinct metamorphosis within each host.

Historically, the term *metamorphosis* was used to describe any change in anatomy during development. Scientists today prefer to restrict the word to major physical changes that occur over a short time. A second, often overlapping, definition encompasses a striking shift from childhood habits and habitats to those of adults. A planktonic larva may feed on single cells of algae, while its adult form preys on other animals. The worm scientist Maslakova describes this transition as, "You go from vegetarianism, picking peas, to eating steak—hunting it down and gulping it whole."[3]

A sudden exchange of one body for another is advantageous when trading habitats. A lengthy remodeling time would not bode well for survival in either place. Ideally, the animal retains as much functionality as possible in the larval body while constructing the necessary adult parts. A simple example is tadpoles growing frog legs long before they're ready to hop out of the water.

Many other baby animals build their adult structures not as external add-ons to their larval forms but sequestered within them. Sea stars and sea urchins achieve a seemingly sudden transformation by getting their new costumes lined up and ready to go underneath the old ones. It can take them weeks to grow these "early juveniles" that allow them to hit the ground running at the moment of metamorphosis. This kind of metamorphosis has evolved many times in many different animal groups, including insects. Larval insects like maggots and caterpillars would be at a disadvantage if they grew wing buds on the outside of their bodies. Such delicate structures would get stuck and torn as the hungry babies work their way through rotting fruit and corpses. So they often grow wings or wing precursors tightly folded under their skin, waiting until the time is right to emerge.

This process is similar to embryogenesis, as many of the same developmental tool kit genes are used to produce both larval and adult forms. But these new bodies or body parts get their own special terminology. In echinoderms, the tiny urchin built inside a pluteus is called a "rudiment." In insects, the adult wings and gonads are formed as "imaginal disks." Imaginal disks have also been found in ribbon worms and are probably present in other invertebrates as well. "It's from *imago* [the scientific term for an adult insect], not *imagination*, but it has the feel of a double entendre," says Nate Morehouse, an entomologist. "Imagine growing your adult eyeballs inside your chest as a child, next to your lungs, and then migrating them up to your head."[4]

Horrifying? Several scientists I spoke to would say so, though not without a certain macabre glee. One even compared the metamorphosing juvenile to Xenomorphs from the movie *Alien*.

Compounding the disturbing visual of an adult bursting free from its larval body is the additional trauma of the animal directly consuming its larval remains. Many snail veligers eat their velum once they no longer need it to swim and collect food. Then there are the solemyid clams, many of which have no guts as adults and rely entirely on nutrition produced by their symbionts. These clams eat a single meal during their lives: their own cells during metamorphosis, to ingest the bacteria that will support them for the rest of their lives.

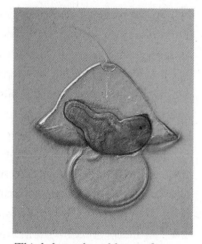

This helmet-shaped larva of a ribbon worm bears cilia arranged rather like those of a trochophore— as well as a growing juvenile worm that will soon burst free and consume its larval body.

In ribbon worms, larvae and juveniles share the same gut. The young worm grows inside the larva, surrounding the digestive tract. During metamorphosis, it consumes the larval flesh even as it emerges from it, "like pulling a sweater over your head and eating it," says Maslakova. This type of change is called *catastrophic metamorphosis*, and it does indeed appear to be a catastrophe. "Often students catch something like this in the plankton and have it on the slide and they'll be like, *Oh, what's happening, oh, it's dying!* And what's happening is metamorphosis."[5]

Metamorphosis is neither death nor birth, but its similarities to both capture our imagination.

Two bodies, one individual

Rebirth, reboot, regeneration—all these words can describe metamorphosis. It is what happens when an animal builds a second body from the same genome. Remodeling, rearranging, breaking down, and rebuilding can offer a second chance at life, an opportunity to fix or discard anything that went wrong the first time around. Scientists have found that tadpoles with facial deformities can often reorganize bone, muscle, and skin during metamorphosis to produce a normal frog head.[6]

Weiss refers to animals with complete metamorphosis as "going through development twice." It's this second chance that she thinks "makes caterpillars and butterflies such an appealing metaphor for people for rebirth, transformation, ascension." Development from a fertilized egg to a larva is remarkable in itself, but even though we all went through that process, it's difficult for us now to identify with a blastula. By contrast, it's fairly easy to identify with a caterpillar. "It's walking around, it's eating its leaves, it's pooping, it's doing all of its caterpillar stuff," says Weiss. "And then that individual undergoes this major transformation, and we get a whole second development where one can think that it is remaking itself. That caterpillar changes into something that seems very different, but it's still the same organism. So the roots of what we become are present in who we are, and things that we learned before can stay with us after."[7]

The fact that memories can be carried across metamorphosis was one of Weiss's most striking discoveries. She and her collaborators trained caterpillars of different ages to avoid a particular smell, then tested them before and after metamorphosis to see how long they remembered their lesson. Smell was the sense of choice because it's the most informative to a caterpillar; their simple baby eyes don't resolve images the way the compound eyes of adult insects do. The scent Weiss offered her caterpillars was ethyl acetate, which we humans associate with nail polish remover, and she conducted a "slightly mean" experiment by giving them an electric shock along with the smell. When young caterpillars in their third instar were trained, they lost the memory by the time they metamorphosed, showing no inclination to avoid ethyl acetate. However, when the oldest caterpillars in their fifth instar were given this training, then allowed to metamorphose and tested again, they remembered to avoid the smell as adult moths.

The explanation for this difference in outcome probably lies in how brains change over time. "What we know about brain development in insects—" Weiss stops herself as she's explaining this to me and rephrases. "What we know about brain development in *Drosophila*."[8] After all, development has been studied in relatively few insects. Most of our knowledge comes from the fruit flies that are ubiquitous in laboratories around the world. Over the course of their lives, fruit fly brains possess three lobes. One is formed early in larva-hood and breaks down during metamorphosis. A second is formed in older larvae and retained through adulthood. The third is formed in the pupa, during metamorphosis. If moths turn out to have similar three-lobe brain development, Weiss's results suggest that the youngest larvae encoded their memory in the first lobe, which did not pass through to adulthood, while the older larvae kept their avoid-the-smell memory in the second lobe, which lasted the rest of their lives.

Adult fruit flies, moths, and other insects may benefit from holding on to their childhood experiences because their adult environment is an expansion of, rather than a complete departure from, their environment as babies. Although flight gives them access to parts of the world they never saw before, they still encounter similar food sources and predators. By contrast, for many marine invertebrates the change in environment at metamorphosis is complete, and echinoderms in particular undergo such catastrophic metamorphosis that their larval nervous system is completely destroyed. In these species, it seems unlikely that any kind of learning or memory could be carried over to adulthood. They're now exposed to an entirely different suite of foods and threats, and might as well start over from scratch, brain-wise. However, other factors can cross this boundary—notably, nutrients. When Emlet discovered that sea urchins could bring some of their yolk across metamorphosis, using it to grow faster and survive better as juveniles, he says, "I was totally blown away."[9]

When it comes to actual food, insects can't make much use of it across the larva-adult divide. Fragments from a caterpillar's leafy diet occasionally pass through pupation, but unlike yolk, they can't be digested by the adult body. Morehouse refers to these bites of leaf as "caterpillar keepsakes marooned in the stomach." Nevertheless, the chemical products of digestion present in the pupa can have an astonishingly strong effect on adults, as in the case of the European map butterfly.

This species goes through two generations in a year. One round of caterpillars pupates in the fall and waits until spring to emerge as bright orange butterflies. These spring butterflies mate and lay eggs, and their caterpillars pupate quickly to produce a round of black-and-white summer butterflies. The children of these summer butterflies again overwinter as pupae. The colors and patterns of the spring and summer adults are so different that they were first described as different species. Once their true life cycle was understood, scientists began coming up with hypotheses about why each pattern might be more advantageous at the time of year it appears. But no one could find evidence to support any of these ideas.

Eventually, Morehouse discovered that map butterfly caterpillars eat plants full of tryptophan, a toxic compound that has to be broken down so as not to cause damage. The products of deconstructed tryptophan circulate through the animal's body during its pupal phase, and when this phase lasts a long time, they build up to produce orange pigments. When pupation is brief, there's no time to turn orange, and wing coloration defaults to black and white.[10]

This case highlights how very little we know about the process of pupation, which researchers often refer to as a "black box."[11] It is the most vulnerable part of the insect life cycle, when they're even less able to run away or defend themselves than caterpillars are, so selection has been very strong for pupae to be difficult to find. Before converting their mobile body into a stationary pupa, caterpillars of different species take different approaches to concealment. Some go underground, others burrow into wood, and for many we have no idea. Our uncertainty about the habitat needs of pupae can hinder conservation efforts, a point that Weiss illustrates with the unexpected tragedy of the butterfly garden.

"Everybody likes butterflies, so they plant pretty flowers, and the butterflies come and visit the flowers. And it's fun to look at the pretty flowers, and fun to look at the pretty butterflies. The [gardeners] are excited about that, and they read a little more and they learn that not only do butterflies need flowers to feed from, but they need plants to lay eggs on, so they plant those plants and they have a nice garden that supports eggs and larvae and adults. And it's so fun to look at, and they feel very good about the whole thing. They really want to be good butterfly stewards, and so they prune back the old plants and rake up the leaves and make sure everything is all ship-shape for the winter, so they're ready for action for the next year. And because we know very little about the pupal phase—for a bunch of butterflies we don't even know where they pupate—and a lot of them pupate in the leaf litter, or in the soil, or in old stems, they can actually be undoing all of the good work that they have done." Such a garden attracts butterflies, encourages them to lay eggs, and feeds the caterpillars, but either provides no pupating habitat or actively destroys it. This turns a "source," a location that produces enough butterflies to seed other locations, into a "sink," or a location that drains members of the population without replacement.[12]

Both small-scale and large-scale human activities impact the mysterious pupal phase. Pupae are especially sensitive to effects of climate change on temperature. For many insects, cold weather triggers a kind of suspended animation, with pupae putting the brakes on development until springtime cues them to pick the process back up. Meanwhile, the aptly named winter moth is active in the cold and pupates over the summer instead. As springs get warmer, winter moth caterpillars that would once have been eaten by the chicks of great tits might begin pupating before the chicks hatch—lucky for the caterpillars, less so for the hungry chicks. Flexible reproduction in the tits, however, has allowed them to adapt their incubation habits so their babies hatch earlier. Thus, although the critical connections that babies weave between species and between habitats are being stretched and sometimes snapped, the flexibility of reproduction and development offers a key to adaptation and survival in a changing world.

Chemicals in the environment, especially hormone-mimicking chemicals, can also exert a strong influence on insect development. Many plants have evolved to produce hormone-like compounds that act as defenses again herbivorous larval insects. Balsam fir produces a juvenile hormone that causes larvae to continue developing as caterpillars and die without ever metamorphosing.[13] Other plants go the opposite route and induce premature metamorphosis, creating sterile adults that can't make more plant-chomping offspring.[14] Without meaning to, we humans have also filled the environment with hormone mimics, like those endocrine disruptors we met in chapter 4. The compounds produced by plants have evolved on a time scale that insect populations can adapt to; the compounds produced by human industry, not so much.

Most of the endocrine disruptors we're leaking all over the place were not developed for the purpose of messing with development. However, there are cases where we humans, like balsam fir trees, could benefit from specifically targeting the metamorphosis of troublesome organisms. When aquatic invertebrates, whose metamorphosis is typically coupled with the selection of their adult habitat, settle on boat hulls, piers, and other underwater structures, the impacts are both economic and environmental—and surprisingly far-reaching. Barnacles and worms encrusting a boat may seem like superficial nuisances, but our global economy depends heavily on shipping. These "biofouling communities" slow ship traffic, hiking fuel expenses and greenhouse gas emissions. And although the foulers themselves are post-metamorphic adults who will remain stuck in place, they can release eggs and larvae anywhere the boat goes, contributing to the problem of invasive species. Research on hormones and metamorphosis may be able to produce anti-fouling treatments to mitigate these problems—although then, of course, we'll be putting more chemicals into the environment.

Settling down

It snowed when I went to Friday Harbor for my final reporting trip. I walked through white woods to visit the Strathmanns—last house on the lane, overlooking the water. My own footprints were the only marks in the fresh powder. Richard and Megumi welcomed me like the grandparents they are, fussing about my wet layers and ushering me into the warmest room.

Over fried rice and brownie cake, they told stories of embryos and larvae, friends and mentors, students and classes. The most astonishing to me was the tale of the Oregon hairy triton larvae. These sea snail veligers are "great big chonkers," according to Megumi, who reared a clutch of them for four years as swimming larvae. *Four years*! She tried to induce metamorphosis by all the usual techniques, to no avail.

This lengthy larval period is hardly standard for the species—when a colleague visiting from California collected a wild triton larva, it settled the very next day. "To get them to live for four years," says Richard, "Megumi was grooming the shells with a camel-hair brush and dental picks," because algae kept growing on them. Finally, she said, "I threw in some rocks that had been dredged up." That did the trick.[15]

The couple mused that perhaps these larvae had become habituated to their glass container and needed a drastic change of environment to stimulate their own change in anatomy and behavior. Whatever the cause, the extended larval life didn't negatively impact their adult survival. "It was as if the delay of metamorphosis was a chunk added to their lifetime," observed Richard. Now, twenty years later, the lab still houses both the colleague's sea snail and one of Megumi's.

The triton larvae, like most other planktonic marine larvae, had to link their metamorphosis to a related but distinct phenomenon: settlement. As they swapped one body for another, they also needed to select the right adult environment. For some animals, like barnacles and oysters, the choice is immediate and irrevocable. For others, including snails like the hairy triton, they'll still have limited mobility once they settle down. But it will be on the order of meters, not miles. The long-distance three-dimensional possibilities of swimming are exchanged for the small-scale two-dimensional range of crawling.

Finding a suitable place to settle is a crucial moment in any animal's life, especially those aiming for a very particular destination. Some larvae venture far afield, then seek out and return to their habitat of birth. Others must aim for a specific adult environment that they themselves have never known, since they were laid as eggs in a distinct nursery habitat. Both these scenarios constitute migration, rather than dispersal, like the flights of adult monarch butterflies. When necessary, babies are quite capable of long-distance migration. Eel larvae travel thousands of miles from the ocean back to the rivers where they will mature. On Christmas Island, baby red crabs settle out of the water on the beaches and then take days to march from the sea to the inland plateau where they will live as adults.

If tiny zoea are swept around at sea as they grow, how do they find their way back to the shores of Christmas Island? This is a puzzle not only for red crabs, but for any planktonic larvae that require specific habitats as adults. Chris Lowe, a developmental biologist at my alma mater of Hopkins Marine Station, puzzles over the case of acorn worms, which can settle and mature only in shallow inlets. "How do you get away with a two-month larval dispersing phase," he wonders, and still find your way to an appropriate home? One day, at the end of a research trip in Moorea during which Lowe hadn't been able to collect any acorn worms, he went out snorkeling and found himself surrounded by hundreds of their larvae. He collected many of them, and by the time he got them home, they'd all metamorphosed into juvenile acorn worms. Clearly they were on their way toward shore, ready to settle. But the next year when he returned, he found no larvae. How did it

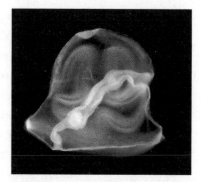

This larval acorn worm bears an uncanny resemblance to a fictional villain from a galaxy far, far away. However, what appears to be a whitish ridge denoting the edge of the "helmet" where it overlaps the "face mask" is actually the baby's gut. Its mouth is open on the upper right; its anus is located on the lower left.

happen that all those larvae were right there, right then?

Just as larval survival is not a matter of pure chance but instead influenced by the larvae themselves, so is settlement. Larvae can sense their environment, seek input, and modify their actions to facilitate the discovery of new habitat or the return to a home range. "Larvae have behavior," says Lowe. "Every time at Hopkins I talk about larvae, they think about hapless bags of fluid that get buffeted around by ocean currents and end up where they end up. [But] they have pretty complicated behavior."[16] The problem is that it's virtually impossible to observe the behavior of such tiny organisms in the wild. A few exceptionally large larvae, like those of sea squirts, have been followed around by scientists who noticed that they used their swimming ability to avoid dispersing far from good habitat. Many more types of larvae can be studied in the lab, helping us to understand how they identify the right landing spot. Sea urchin larvae exposed to turbulence that simulates the crashing waves characteristic of adult urchin habitat promptly prepare to metamor-phose.[17] Both coral and fish larvae, amazingly, track the *sounds* of a reef to find a good place to settle.[18, 19] Throughout the seas, however, the single most common source of "time-to-settle" cues is bacteria.

Meeting microbial neighbors

The marine biologist and embryologist Michael Hadfield says, "When the first pelagic larva wanted to settle, it could not find a clean surface." Bacteria, after all, had evolved almost as soon as Earth got cool and wet enough to be habitable—nearly four billion years ago. Animals have only been around for the most recent one billion of those years. By the time early worms and snails and fish were looking for places to live, bacteria had already colonized every location. The sea is especially rich with a diversity of bacterial and viral life, the complexity of which we're only beginning to grasp, since most of these organisms still can't be cultured in a laboratory. In the saltwater habitat that dominates almost three quarters of our planet, every ecosystem is built by the larvae that settle in it, and the majority of these larvae respond to bacterial cues.[20]

"When I was starting grad school, the whole dogma was that larvae were so abundant that it was just luck that some of them fell in the right place," says Hadfield. But careful experiments had begun to show that certain larvae were more choosy than fortu-nate. Hadfield wanted to explore the chemical nuts and bolts of this choosiness. Was there a replicable recipe for metamorphosis and settlement? He started off with *Phestilla*, a sea slug that lives on coral in both senses of the word—it dwells on coral, and it eats coral. Sea slugs, also called nudibranchs, look far more exciting than the garden slugs most of us are familiar with. They resemble elon-gated, rubbery fireworks, often covered with colorful frills and gills that wave in the water. Like sea snails, sea slugs produce veliger lar-vae, and their veligers grow a baby shell that is lost at metamorpho-sis. This shell turned out to be crucial for Hadfield's experiments.

To observe *Phestilla* larvae, he needed to hold them still by attaching them to the tip of a pin. He and his colleagues tried to connect larva to pin with every kind of glue at the hobby store, but the larvae either fell off or died. One morning while shaving, Hadfield remembered seeing the shells of larvae get stuck sometimes at the water surface, because they have water-repellent properties almost like droplets of oil. He wondered if he could make use of this feature. "I dug around way under the bathroom sink and found a jar of Vaseline and took it to the lab. I took one of my nice silver pins, and I poked it in the Vaseline. I had a bunch of larvae in a little petri dish and I chased them around, and when I touched one it was like a magnet."[21] The water-repellent nature of the Vaseline matched that of the shell, keeping the two stuck together even while wet.

With this "shaving brilliance" to help him observe the effects of different cues on individual larvae, Hadfield demonstrated that the settlement of *Phestilla* was far from lucky or accidental—it depended on one specific chemical from one specific coral. Despite years of collaboration with chemists, however, the molecular structure of the settlement cue has never been elucidated. It simply can't be collected in large enough quantities. "It is so powerful a molecule, mind you, that if you had a visible quantity, you'd make every nudibranch in the world metamorphose," says Hadfield. "We know a lot about it, without knowing what it is."

He'd proved that larvae could detect and respond to signals in their environment, selecting a place to settle based on their compatibility with its current inhabitants. From sea slugs he moved on to a type of bristle worm named *Hydroides* that grows a hard tube for itself after settlement, and found that bacteria induce its metamorphosis. As Hadfield and other researchers branched out into a broad diversity of species, they discovered bacteria-induced metamorphosis time and again—in sponges, in clams, and in corals themselves.

Hadfield recalls when scientists used to think that larvae metamorphosed because their time had come, that they came "raining down and those that landed on fertile soil" were the successful ones. But evidence is mounting that marine larvae rarely have a rigid countdown clock for metamorphosis, as dramatically illustrated by Megumi's four-year triton larvae. "I think that's one of the most important things we've ever found," says Hadfield. "That the larval period is such a rubber band."[22] The paradigm has shifted, so arguments now focus on what the cue is rather than on whether or not there is a cue. For years scientists thought that coral larvae were cued by algae to metamorphose and settle, until new research indicated that bacteria living on the algae provide the real cue. Here larvae once again act as links: between the amorphous world of plankton and the highly structured communities on rocky shores, between minuscule microbes and the city-size sprawl of coral reefs, oyster beds, and kelp forests.

Although *Phestilla*'s cue remains a chemical enigma, that of the tube worm *Hydroides*

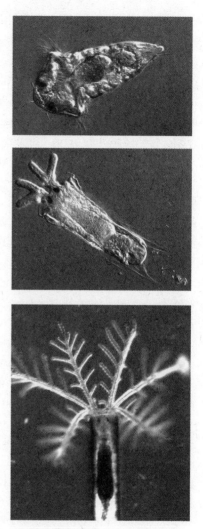

While swimming freely in the sea, a larval tube worm develops from a trochophore into a setiger (a). When it encounters the right cue, it settles to a hard surface to begin growing a tube and feeding tentacles (b). Within two weeks, its tube has taken shape and its tentacles have proliferated (c). The cute eyes remain throughout.

has proven more tractable. The molecular biologist Nick Shikuma at San Diego State University figured out the bacterial process that induces these worms to settle, and the details are, shall we say, unsettling. When worm larvae come into contact with the right bacteria, syringe-like structures in the bacteria inject molecules directly into larval cells.[23] These molecules, like the coral compounds that induce metamorphosis in *Phestilla*, are very powerful. In high concentrations, they simply kill the worms. Experimental application of the molecules to other larvae, like baby insects and jellies, is also fatal.

Why are tube worms uniquely suited to taking small doses of this deadly chemical as a cue for metamorphosis? It might give them an opportunity for early settlement—if they can recognize as a signal something that kills other animals, that gives them a leg up on free real estate. *Hydroides* do tend to be the first animals to colonize or recolonize any available surface. As one of Shikuma's students told me, if you see a boat covered with tube worms, it's a boat that's been recently scraped clean of uninvited passengers. A boat hull decorated with barnacles has gone far longer between cleanings.

What about the bacteria? Did the production of this molecule evolve to bump off their animal competition so they could keep the surface for themselves? Or do the bacteria benefit from recruiting worms? One of the latest hypotheses is that after tube worms settle on a surface, bacteria have the opportunity to colonize these juvenile worms, finding a new home as internal gut symbionts. To fully explore such a possibility would require maintaining juvenile worms in the laboratory and culturing them to adulthood. Unfortunately, raising juveniles is trickier than spawning adults, trickier than fertilization, even trickier than feeding and growing larvae.

The contrast reminds me of a common lament from human parents, that adolescence turns their familiar young children into mysterious and challenging teenagers. This new phase of youth has new needs, but what are they, and why are they so difficult to provide?

10

JUVENILES

Neither One Thing nor Another

We have to-morrow
Bright before us

Like a flame
Yesterday, a night-gone thing
A sun-down name

And dawn to-day
Broad arch above the road we came,
We march.

—Langston Hughes, "Youth"[1]

A hermit crab, like many other crustaceans, first builds itself into a cyclopean nauplius while still in its egg, then metamorphoses into a second larval form called a *zoea*. The spiny, big-eyed zoea hatches, drifts in the plankton, and metamorphoses into a *megalopa*, a third larval form that looks like a tiny crab-lobster hybrid. The megalopa commences the hunt for a good shell to live in, a shell that must of necessity have been left behind by a dead snail. It has to be the right size, and they have preferences about shape, too, choosing rarer shells over more common ones.[2] A hermit crab's transition from larva to juvenile depends on precisely the right death, and so, as these animals connect sea to shore, water to land, they also connect death to life, endings to beginnings.

Such is the glory and burden of youth. In school the lessons of the past are impressed upon us, and our promise for the future is constantly under scrutiny. Career counselors and college applications ask: What will you make of yourself, what will you make of the *world*? Our adolescence and young adulthood are rife with complex challenges, just like those of the itsy-bitsy spiders and gangly condor fledglings we'll spend time with in this chapter.

The juvenile phase of any animal's life cycle presents quandaries faced by neither babies nor adults. Juvenile sea stars, urchins, snails, and oysters are still incredibly tiny right after metamorphosis and face quite different environments compared to adults. A tube worm the size of an eyelash, a sea urchin like the head of a pin—they are adult-shaped, but not yet adult-size. We humans face a different mismatch: Our bodies become adult-shaped and adult-size up to ten years before our rational brain is fully developed. What can we learn about adolescence from studying the "in-between time" of other animals?

The mystery of juvenile lifestyles

The puzzle of post-metamorphic juveniles is really two puzzles: where they are and how they survive. For many species, such as the red sea cucumber in California, both larvae and adults can be collected in the wild, but no one has ever been able to find newly settled juveniles. People fish these animals for food, and fishery managers would really like to be able to assess all parts of the life cycle in order to figure out what a sustainable harvest would look like. We know that when the distinct ecological needs of intermediate life stages are ignored, the health of a fished stock can be threatened. For example, baby salmon need gravelly streambeds and adult salmon need oceanic feeding grounds, but juvenile salmon need pools for overwintering. Scientists discovered that efforts in the 1960s and 70s to "clean up" streams by removing fallen logs and branches had cleared away these necessary habitats, contributing to the decline of salmon populations.[3]

It can be daunting to think that we might need to hunt down juveniles species by species in order to manage them properly. However, although this information would be incredibly useful, we can generalize about juvenile habitats in certain groups. For example, a wide variety of juvenile marine fish seem to settle from the plankton into one of three "nursery" types: seagrasses, mangroves, and kelp forests. Marine parks can be designed to emphasize protection of these nurseries—which is also likely to preserve habitat for invertebrate juveniles, like sea cucumbers. This ecosystem-based management approach is an integrative way to proceed with conservation even in the absence of knowledge about many species' life cycles, and it seems to be an effective one.

Unfortunately, there are species for which we know exactly where the juveniles are, and still the question of their survival remains vexing. Oyster juveniles are easy to find because they like to settle on adult oyster shells. For hundreds of years, managers of oyster fisheries around the world have been placing plates covered with old oyster shells in the water to encourage settlement. These animals, however, perfectly illustrate the gap between *settlement* and *recruitment*. A settled juvenile has chosen a substrate and undergone metamorphosis. A recruit, by contrast, has grown large and capable of reproduction. It doesn't matter to the population or to the fishery how many oysters settle on plates, if they die before they're big enough to reproduce.

The threats they face, and the consequent likelihood of recruitment, seem to vary tremendously from place to place. On the Pacific coast of North America, a couple hundred miles is enough to make the difference between reliably successful recruitment (Southern California) and very rare recruitment (Central California). From time to time, enough oysters reach adulthood to keep the Central population going, but these recruits are few and far between. Whatever environmental factor drives the variability in juvenile survival has yet to be identified.[4]

When I ask Richard Emlet the admittedly leading question of whether juveniles are the least understood part of most life cycles, he claps his hands and shouts, "Yes, yes! I've thought that since I was on the rocks in Panama."[5] He spent his student days in tide pools, and he's never left this habitat behind. From all his hours exploring cracks and crevices, he knows that juveniles of most species are elusive to downright invisible. Often the very smallest individuals that can be found are still, by all growth rate calculations, several weeks post-settlement. Where do the juveniles hide during their first hours and days?

Emlet marvels that it's even possible for the adult shape of a chiton, a snail, or a sea urchin to function at such a minute scale. Larvae, after all, are thought to have evolved at least partially because different shapes operate more effectively at different sizes. But if the adult body shape works at the near-microscopic level of some juveniles, that would constitute a counterexample to this theory. "What is the smallest viable sea urchin?" asks Emlet. "That could be answered in a laboratory dish, but it won't necessarily be the ecological answer." Rearing juveniles in an aquarium can provide only limited information about how they would live and cope with their environment in the wild.[6]

The same limitation is also true for larvae, Amy Moran points out, and yet scientists still bumble along learning the best they can from larvae in the lab. Larval biologists who raise their study organisms in captivity are well aware that they're not exactly re-creating an oceanic environment. "Mostly we feed them to the gills with some algae that came from Tahiti that they'd never encounter in nature, and we keep them at very high density," says Moran. A huge amount of valuable information has been gathered this way, while scientists simultaneously attempt to push larval research into the field with new technologies. Meanwhile, researchers have remained leery of trying to rear juveniles in an "unnatural" laboratory environment. But what even is their

natural environment? "We admit we have no idea where the juveniles end up," says Moran.[7]

Numerous biologists in recent years have begun to abandon
this historical reticence. Acknowledging the caveats and limitations, these careful researchers are gathering valuable insights into
long-standing mysteries.

Sunflower juveniles: the stars of the show

The CRISPR technology that Hamdoun's lab uses to produce
bespoke sea urchins is a sophisticated molecular biology technique.
DNA must be cut in precise locations, then new genes and markers
must be carefully spliced in. "You would think that would be the
focus, but actually that stuff is super easy, and it's the fish farming
that we're spending all our time on," says Hamdoun. "What do
you feed an urchin that's one millimeter in diameter? How do you
make it happy? It looks like a little grain of sand."

Plutei, the larvae of sea urchins, are easily raised in captivity on a
diet of single-celled planktonic algae, which they collect with their
cilia. Adult urchins retain this herbivorous habit. In fact, they're
famous for devouring entire kelp forests if their populations are
unchecked by predators. But a fresh kelp frond is dauntingly thick
and rubbery for a diminutive urchin to chomp on. Juveniles need
some kind of transitional food, like human babies moving from
milk to solids. Traditionally, human parents have partially chewed
food for their children, and Hamdoun arrived at a similar solution
for juvenile urchins.

"I was watching a show on Netflix about fermented food, so I
thought, let's decompose the kelp," he says.[8] Kelp decomposes with
the help of microscopic algae related to the planktonic algae eaten
by plutei. These algae grow in a film over the rotting kelp, a film
that is easy for juvenile urchins to scrape off and eat. The kelp as
it falls apart is also softer and easier to bite than a fresh kelp frond;
this combination of algal film and decomposing kelp provides the

crucial intermediate dietary step. Like the insect larvae that hasten the breakdown of feces and corpses on land, these marine larvae are linking cycles of death to cycles of life.

Many other echinoderms, however, must make a more significant dietary leap—from herbivory to carnivory. This is the challenge faced by the ecologist Jason Hodin as he works to rear sunflower stars from embryos to adulthood (see insert, photos 14 and 15). Hodin and I overlapped as graduate students at Hopkins, and I was excited to visit his current research lab in Friday Harbor when I made my final book reporting trip in the winter of 2021 to 2022. His focus on sunflower stars comes at a time when many species of starfish have been devastated by a mysterious wasting disease, losing skin and arms and eventually melting into death. Although minor outbreaks were known in the 1980s, sea star wasting disease began truly ravaging both coasts of the US in 2013, and scientists reported cases from Europe in 2022. The lethal symptoms have been variously ascribed to a viral infection, a bacterial infection, or an overall imbalance in the microbiome. With its cause still unidentified, we don't know how to limit its spread or whether it could affect other kinds of animals. The loss of starfish alone has significant ecological impacts. Sunflower stars, in particular, are major predators of sea urchins. When sunflower stars on the west coast of North America were decimated by wasting disease in 2013, sea urchin populations ballooned, and kelp forests suffered huge losses. Hodin's work with sunflower stars aims to understand and, ideally, ameliorate the situation.

Sea stars and sea urchins are both classic participants in embryology studies, and spawning them to make embryos is routine. We'd done it plenty of times when I was a student in Strathmann's class at Friday Harbor. But, like most embryologists and larval biologists, we always tossed our metamorphosed juveniles back into the ocean. Hodin is the first to successfully raise sunflower stars all the way from fertilization to two-year-olds.

He shows me an aquarium full of animals about the size of my palm with a practiced pun: "These are the stars of the show."[9] Prior to his work, when people saw a sunflower star of this size in the wild, they could only guess at the animal's age. Estimates ranged up to ten years. Now we know that it's possible for them to grow to this size in two years. Not as speedy as the growth of a black soldier fly larvae, but still nothing to sneeze at—and they *are* getting considerably larger than a fly.

We go outside so I can meet the adults that provide the eggs and sperm to make each new round of babies. Their tanks are dusted with snow, which astonishes me for a moment until I remember that the beaches themselves have been snowed on. There's no reason to believe that sunflower stars aren't adapted to cope with the occasional cold snap. Each parent star is the size of a stop sign, so big it would fill my arms if I picked one up. They sprawl over a substrate of picked-clean mussel shells, both beautiful and gruesome. Their names are inspired by their appearance: Van Gogh is especially colorful, Prince is purple, Ulva (the scientific name for sea lettuce) has an almost greenish fuzz, Clooney is going gray.

As we gaze at these bright giants, Hodin tells me wistfully, "If you'd come two months ago, I could have told you we've had no wasting disease." But then one of his adults lost an arm and succumbed shortly after that. After sixteen days of hoping that it was an isolated incident, the disease spread, and in the end his lab lost eight adults. However, the remaining animals have been healthy for weeks now.

I ask about recent cases of wasting disease in the wild, around the island. "There's been an uptick. Who knows? Maybe this is the 'omicron' of sea star wasting disease," says Hodin, making a grim comparison to the most recent wave of coronavirus at the time. "I'm so frustrated not to know what it is. We'd like to be able to test the water." But no one knows whether it's caused by bacteria, a virus, or something completely different, so what can you test for?

He points out that it must be something really flexible, because it can affect all kinds of sea star—a group of animals that's more genetically diverse than all the world's mammals. "Imagine the next pandemic killing people's pets. Killing horses. Whales getting sick," he says.[10] It gives me chills.

Hodin himself isn't trying to figure out the cause of wasting disease; rather, he's interested in figuring out the life cycle of sunflower stars. Perhaps information that he gains in the lab can be applied in the wild. Perhaps sunflower stars can be raised in large numbers to bolster wild populations, like white abalone and condors. He has already discovered one vulnerability of juvenile sunflower stars, along with a solution for it.

The labs at Friday Harbor pump seawater in from the ocean, and it carries plenty of microscopic guests. One of these, an opportunistic brown alga that covers tanks in brown fuzz, turned out to be toxic to the juvenile sea stars. Hodin had to overcome his embryology-trained aversion to bleach, and now he and his team routinely bleach their holding containers, then thoroughly rinse off the bleach before putting animals in them. Although bleach was always a big no-no in the "embryo-clean" environments where Hodin and I were trained, it's commonly used in aquaculture, or fish farming, facilities. His team can wash the containers so well that the most sensitive bleach test kits will read zero—although he's found that the human nose is even more sensitive and can still smell the bleach. "Of course, it's hard to smell *now*," he says dryly, gesturing to the face masks he and I are both wearing.

As an additional layer of protection against the toxic algae, the researchers have taken to growing "probiotic algae" of a kind that won't bother the juveniles. These beneficial algae compete with the toxic species for space and may even be grazed by the stars to a certain extent. However, juvenile sunflowers stars are not, for the most part, algae eaters. They are predators. The smallest ones will eagerly devour each other if given half a chance.

"Each other" seems to be one of the only foods these tiny juveniles will reliably eat. So far, the best alternative to siblicide that Hodin has found is a diet of tiny baby sea urchins. This illuminates how early life stages, despite their inconspicuous nature, have the potential to exert outsize impacts on the ecosystem. While sunflower stars are considered valuable biological control for urchins, the focus tends to be on adult stars eating adult urchins. If the juveniles of one are also eating the juveniles of the other, that interaction could be equally or even more important.

Curiously, Hodin's lab has found that some sunflower juveniles survive, but don't grow. They seem able to live for months without getting larger. And they're not the only species with a prolonged ability to hang around as juveniles waiting for better conditions—crown-of-thorns starfish do it as well.

The crown-of-thorns starfish, like the sunflower star, is a voracious predator. However, the role it plays in its ecosystem is more similar to that of the sea urchin, which can threaten entire kelp forests. Crown-of-thorns are coral eaters, and when their populations periodically explode in abundance, they inflict enormous damage on coral reefs.

Unlike sunflower stars, crown-of-thorns don't switch from eating algae to eating coral at the same time as their metamorphosis from larva to juvenile. Their juvenile digestive system isn't yet capable of processing the available nutrients in coral, but even more important, they're still small enough that a coral colony could sting them to death. Juvenile sunflower stars can find juvenile sea urchins to eat, but crown-of-thorns stars don't have a similar option. It's a whole coral colony, or no coral at all.

So when they settle to the seafloor, juvenile crown-of-thorns remain algae eaters, sometimes for an astonishing length of time. Maria Byrne, the Australian biologist who studied gutless urchin babies, once kept a bumper crop of crown-of-thorns juveniles alive on a vegetarian diet for six years.[11] Crown-of-thorns stand out among sea stars for being obligate algal grazers as juveniles, but Byrne thinks supplementary grazing is probably common to many other species, like the sunflower juveniles that snack on their algal probiotics.

"The juveniles are a big black box," she says. "We put juveniles in the 'hard' basket. They're still hard, but we need to recognize this is the missing link in the big puzzle. We cannot understand adult populations until we understand what happens at the juvenile stage."[12] Byrne thinks many marine invertebrates could have a juvenile waiting stage, its length determined by resource availability and the size of the existing adult population—similar to the flexible larval duration we saw illustrated by baby triton snails. It could be more common than not for young animals to postpone maturity until the time is ripe.

Spiderlings and juvenile vision

Feeding herbivorous marine larvae in a laboratory setting is relatively easy, as scientists have their techniques for culturing algae down pat. A few beakers of algae, some sunlight, and there's plenty to go around. This is one of the rare situations where terrestrial biologists face a greater challenge. Herbivorous land larvae like caterpillars go through enough roughage that you can't keep up by growing it in the lab. Morehouse says he found himself making regular trips to Whole Foods for cases of cabbage and kale to keep his caterpillars content. "There was a betting pool among the cashiers what I was using it for," he says. "And top of the list was kale enemas. I was like 'No, no, no, it's much more normal than that, I'm feeding it to caterpillars!'" He speaks fondly of his thousands of lab larvae. "You could hear the quiet sound of them chewing their plant, en masse."[13]

However, I called Morehouse to talk about a different kind of animal baby: spiderlings, which he describes as "little noodly things." I was interested in these babies because they hatch at minuscule sizes and must immediately cope with an adult-like environment in an adult-shaped body. Looking closely, you can see that their "noodles" are the eight jointed legs of a typical spider, attached to a typical spider body with head and abdomen. Morehouse and his student John Goté wanted to know how these itty-bitty babies use their itty-bitty eyes to hunt prey.

Spiders, like humans, do not have a true larval stage, so they don't get a chance to grow in size first, then build a new body the way caterpillars do. They have to start life with all the important bits already in

Like those of a young human or other mammal, baby spiders' eyes are enormous for their body-size. They have to pack in every photoreceptor they'll ever need for the rest of their lives.

place—including all of the eye's light-sensing cells, called photoreceptors. Both baby humans and baby spiders have enormous eyes relative to their body size in order to pack in the photoreceptors, but this doesn't create the equivalent of adult vision. Newborn humans don't see very well, with the overall shape of the eye and the brain's ability to process its signals still undergoing development. Baby toys are often designed with simple, high-contrast designs to suit the limitations of early vision. Unlike human infants, though, spiders are responsible for feeding themselves from the get-go. How can they find and capture prey with such tiny eyes? To answer this question, Goté collected spiders in the wild. Or, perhaps not quite the wild. He found "massive colonies of spiders" between the greenhouses of local organic farms, from which he was able to acquire plenty of adults that would lay eggs in the laboratory. Once hatching commenced, he found adelphophagy—sibling cannibalism—to be a prompt concern, so he separated the baby spiders as quickly as possible. They were born predators, capable of taking down not only each other but also crickets equal to or even larger than their own size.

The greatest challenge Goté faced was arranging the spiderlings in an ophthalmoscope so that he could study their vision. With slow and delicate work, he found that their eyes are "perfectly optimized" to gather light and detect movement. They could, in essence, do everything that an adult spider eye can do—with the caveat that they need more light to manage it. "Once you dim the lights, that's when the juveniles start to have issues," says Goté.[14] The vision of human infants is also limited in low light conditions, and researchers have found that regular exposure to bright outdoor light helps children's eyes, making them less likely to develop nearsightedness.[15]

Curious about the potentially comparable needs of spiderlings, I ask if they're constrained to hunting at brighter times of day and in brighter habitats compared to adults. Goté doesn't know and says that he would love to study it. But as tricky as it is to manipulate a spiderling in the lab, it's even more daunting to find and observe them in the wild.

Land or sea, field study of most invertebrate juveniles is hampered by their tiny size. It's hard enough to find them, and if you can find them, how can you apply standard research tools to their situation? One Hawaiian limpet lives at the very edge of the seashore, on rocks rarely splashed by waves. Typically, such a hot and dry environment would be a challenge for limpets. To understand how adult limpets cope, scientists would measure air and rock temperatures around the animals, study their internal chemistry, and find out what strategies keep them from cooking and desiccating. But juvenile limpets the size of sand grains make everything about such a study more daunting. How can temperature be measured on such a tiny scale? Is there a microclimate right at the rock surface, relevant only to the littlest organisms? Their bodies are too small to insert probes into, and difficult if not impossible to extract chemicals from.

Although scientists may yet struggle for years with pulling invertebrate juveniles out of the black box, we can get helpful context from species with much larger juveniles that are relatively easy to study in the wild: birds.

Avian adolescence

As Burnett explained to me, researchers have observed pretty much every detail of condor growth from hatching onward. The pink heads of newborns turn black and grow fuzzy feathers as they develop into juveniles. Moving toward adulthood, their skin color shifts again to reddish orange, and, Burnett jokes, "They get bald like me." The classic explanation for the baldness typical of scavenging birds is that it prevents them from getting bits of flesh and entrails caught in their feathers when they stick their heads into a carcass to feed. However, a few years ago, a study on griffon vultures indicated that a more significant benefit may be the ability of bare skin to regulate the birds' body temperature. Without insulating feathers, this part of the body can exchange heat more quickly with the environment. When the vultures get too hot, they can stick out their head and neck to cool off, and when they're cold, they can scrunch it in to stay cozy.[16]

It seems like young condors would benefit from both aspects of a naked head, so why does it take them years to lose their cranial feathers? One possibility is that for condors, a nude noggin is more connected to social behavior. The skin of adult condors flushes and turns pale during their interactions with each other. Young condors interact mainly with their parents and have no need for complex signaling with other members of their species. As they mature, their colorful, feather-free heads allow them to express feelings and communicate, essential to mature condor life. This remains only a hypothesis for now, as condor researchers have been so focused on the species' survival that they haven't had a chance to conduct long-term behavioral studies.

Distinct plumage changes accompany adolescence in many other birds, and, as with condors, we don't always understand why— even in the seemingly obvious case of flamingos. The famously pink

Baby flamingos may look like incomplete, if adorable, adults. But they've adapted to their own niche: fluffy down for warmth; white and gray coloration for camouflage; and short, straight beaks to gulp crop milk from their parents.

coloration of adult flamingos comes from their diet, rich in carotenoids. However, their fancy feathers are more than just a side effect. Flamingos preferentially store carotenoids that are more protective from sun damage, and they also make use of their bright colors as mating signals. Meanwhile, flamingo chicks are born white and quickly darken to gray or brown. It had been thought that their plumage pinks up as soon as they've eaten enough carotenoids. However, a closer look reveals that juveniles retain their dark feathers until they're four to six years old. Evidence suggests that it's advantageous for them to delay looking like adults until they're ready to breed, otherwise adults might act more aggressively toward them.[17] This isn't too different from condors gaining bald heads only when they're ready for more social interaction, and it happens at a similar range of ages.

Like humans, condors can be early or late bloomers. They typically hit their equivalent of puberty at around four to six years, but some don't mature until they're eight. During this time of transition, young condors continue to benefit from parental care.

Instead of bringing food, now the parents of a juvenile condor are introducing it to the flock, teaching it how to interact and find its place in the hierarchy. The confidence and skills to participate in social life are "as important as flying," says Burnett. "They could be a perfect flier, but if they don't know how to work with the flock in a very social, gregarious species, they're done. They're not going to make it."[18]

Condors are hardly the only species in which juveniles rely on their parents even after the early stages of direct provisioning are over. Juvenile scorpions have special feet for holding on to their mom's back, and even after they begin walking around on their own, they'll still return to her periodically for safety. Similarly, a marsupial joey large enough to leave its mother's pouch will continue to visit that haven for quite some time. Male chimps and killer whales both remain dependent on their mothers for support well into their adolescence and even adulthood.

But there's a fierce need for independence, too. When condors are ready to mate, they leave their parents and travel enormous distances. Groups of juveniles begin to fly and scavenge together, and eventually pair off. "I remember when I first left home, I drove three thousand miles from Virginia to California, so I see why these birds do it," says Burnett. "They're like awkward teenagers. You try not to anthropomorphize, but you can't help it."

Describing the development of nonhumans with words like adolescence and puberty, I'm certainly guilty of anthropomorphism, too. But could this way of thinking be bidirectional? In addition to seeing how condors and spiders are similar to us, we can also begin to see how we are similar to them. When we look at the joys and challenges of our own childhood, or that of our offspring, and recognize aspects of development that we share with our fellow animals, we gain a gentle distance from ourselves. The struggle from infancy to maturity is shared across the kingdom, from baboons to bugs.

EMERGENCE

A Cicada Case Study

The cry of the cicada
Gives us no sign
That presently it will die.

—Matsuo Basho, "The Cry of the Cicada,"
translation by William George Aston

I signed the contract to write this book in February 2020. As you can guess, this meant that I wasn't able to travel for research as much as I might have liked. With the exceptions of a visit to the baby sea urchins in San Diego, and to the juvenile sunflower stars in Friday Harbor, I met all the generous scientists and fascinating baby animals that feature in previous chapters through videos, photos, and phone calls. In a way, it felt like we were all embryos inside eggshells. Any environmental input was filtered through protective layers of travel bans and social distancing. Our behaviors and the trajectory of our development were influenced by everything going on in the wider world, but our physical surroundings were severely circumscribed.

Then the spring of 2021 brought a dramatic opportunity.

After fifteen months of a global pandemic, I was ready to crawl out of my skin, climb a tree, and scream. When I learned that trillions of cicadas in the eastern United States were planning to do exactly that, and the spectacle was forecast to begin two weeks after my second COVID vaccine dose, I couldn't book a flight fast enough.

In most of the world, including my home state of California, the summer cry of the cicada is background noise. The insects themselves are hard to spot. But a different kind of cicada lives in the eastern US, emerging en masse every seventeen years to put predators into a food coma and drown out jet engines with mating calls. Humans reactions run the gamut from irritation to terror. The Maryland entomologist Gaye Williams says she's "tired of talking people down off the ceiling." Once she got so sick of fielding a family's worried calls about an outdoor wedding that she told them to serve shrimp dip, so guests wouldn't be able to tell if cicadas had fallen in.

But the unique biological phenomenon has also spawned cicada chasers, bug brains who'll travel hundreds or thousands of miles to witness an emergence. And now I'm one of them.

A childhood as long as our own

It was the life cycle of the cicada that entranced me. Although adults make headlines with their showy abundance, they all die off in a matter of weeks. But before metamorphosing, cicadas have a lengthy youth, comparable in time span to the childhood and adolescence of humans. The cicada life cycle is like the mayfly life cycle multiplied by seventeen. One year of youth becomes seventeen years, one day of maturity becomes seventeen days (give or take a week or two). Also like mayflies, cicadas are born as nymphs, not larvae. Caterpillars, maggots, grubs, and other true insect larvae tend to resemble worms; nymphs are more insectoid in shape. They have six legs, compound eyes, antennae, and of course distinct adaptations to their environments. While mayfly nymphs are born with gills, cicada nymphs are born with digging claws to burrow underground in search of tree roots. When they find one, they latch on and suckle the tree's juices for the next seventeen years, growing and molting and tunneling no more than a few feet from their birthplace. The joke is that their last news from the topside world was the reelection of George W. Bush, but in reality they haven't been so isolated. Cicada nymphs taste the seasons in the changing tree sap, and they sense every degree of temperature shift in the soil.

There are always cicada nymphs underground, waiting. They're organized into broods, based on year of emergence and tracked with Roman numerals. In May 2021, the nymphs of Brood X (pronounced Brood Ten), the largest group of seventeen-year cicadas, began crawling out of the dirt to metamorphose. At this juncture, they trade their digging claws for a pair of wings—and, in males, a sound system to rival any souped-up sports car. Unlike the shy Californian cicadas that silence themselves when I draw near, the East Coast cicadas make no effort to hide their racket from potential threats. Their survival strategy is simply to show up in such numbers that predators can't possibly eat them all. The noise they make has been described as screaming, yelling, drumming, and buzzing. It's been compared to flying saucers and chainsaws. It is the sound of sex, or rather, the sound of foreplay, since they shut up once they get some. Having mated and laid eggs in the branches, adult cicadas will all drop dead by July, decomposing to nourish the very trees that will feed their offspring.

When baby cicadas hatch in the fall, they're so tiny that their abundance doesn't garner much human attention. But it sure attracts the notice of relevant predators. Ants and spiders are ready to snap them up at once. These few weeks of hatching offer a baby buffet for any small hungry carnivores, a hidden boost of energy in the ecosystem that will trickle up into larger predators. The nymphs' only protection is to hide underground, so they don't waste time crawling down the tree—they let go and let gravity work.

First they feed on grass roots, then after a few weeks they graduate to tree roots, usually belonging to the tree in which they were laid, although in a dense wood several trees' roots might be jumbled together. This underground habitat is a shelter not only from predators, but from freezing temperatures. The thick blanket of soil serves as an insulator, keeping the babies at a cozy 56°F (13°C) even when there's snow at the surface. It seems that animal babies have evolved to take advantage of every potential incubator on the planet, from deep-sea hydrothermal vents to deep-dirt tunnel beds.

I'm struck by the contrast between the adult behavior that grabs our attention (they fly, they scream, they screw) and the quiet, diligent larval behavior that determines when they come out. It's the nymphs that keep track of time and temperature. When they're very young, their underground activities are not perfectly synced. Some cicada nymphs molt through early instars before others. But somehow all the nymphs of a brood arrive together at their final instar, and then they all wait exactly four more years beneath the surface. They remind me of embryonic chicks that, even if laid many days apart from each other, can communicate through their eggshells to coordinate hatching at the same time. This chick-to-chick communication is mediated by a combination of vibration and sound (which is, after all, just a kind of vibration), but we're still in the dark about how cicada nymphs coordinate and track time. It happens so slowly that it's virtually impossible to conduct experiments.

However, capricious climate provides us with a natural experiment from time to time, indicating that nymphs track the passage of years with the annual change of sap flow in the trees. Brood XIV was due to emerge in Cincinnati in 2008. But an unusually mild winter in December 2006 and January 2007 caused the maple trees to bud early. Then a February freeze killed these early leaves, and the trees had to do it all over again when spring arrived properly in April. So 2007 saw an early emergence of Brood XIV cicadas, because the false spring had caused them to count the passage of an extra year.[1]

Curiously, nymphs don't always count by ones. If a false spring occurs in the first five years of their lives, their emergence becomes accelerated by four years rather than a single year. Numerous broods of cicadas have demonstrated that an early "extra year" causes them to go through an extra molt as youngsters, which then knocks four years off their time underground. "We think that there's some kind of a switch that turns on a four-year increment," says Gene Kritsky, a cicada expert. Broods of periodical cicadas in the United States all spend either seventeen or thirteen years underground—a difference of four—and when they emerge off-cycle, it's always early or late by either one year or four years. The four-year switch seems to be even more ingrained than the extra single year; periodical cicadas in Fiji emerge on a simple eight-year cycle, and in India on a four-year cycle.

I wasn't able to meet Kritsky in person, as he's based in Cincinnati and the 2021 emergence was centered on Washington, DC. When I tell him over Zoom that I'll be going to see the cicadas, he asks, "So you're going to fly?"

"I'm going to fly," I confirm.

"Oh, you're a brave soul, lassie."[2]

I knew there were ample reasons to be worried about flying in pandemic times, but my concern hovered exclusively around the weather and bug situation. First I worried I'd be too late—if I arrived after most of the cicadas had emerged, I'd miss their metamorphosis. Then a cold snap arrived, and I worried I'd be too early, forced by nonrefundable tickets to return home before the singing cicada spectacle really got going.

The day (and red-eye night) of travel finally arrived. When I completed my emergence from Reagan National Airport and molted out of my PPE, my ears were not assaulted with cicada song. So, not too late. I decided to be relieved, and revel in this new environment where I'd settled, however briefly. I felt a rush of heat in the air, and with it a rush of optimism in my heart. I drove past green leaves and red brick, savoring the palette so different from my home of brown and gold.

I saw my first cicada in the garden of the friend's house where I'd be staying. There's nothing quite like the thrill of recognizing a bit of biology for the first time—a species, a behavior, or an anatomical marvel—and then suddenly beginning to see it everywhere. I glimpsed one empty discarded molt, then dozens more. I spotted a big buzzer zooming across the yard and thought: "Too big for a fly, not the right shape for a butterfly—*it's a cicada!*"

I bent to look at a very still adult cicada, and as I looked, it took off, flying across the garden with more enthusiasm than skill. I watched another crash-land in a hosta plant, its body wedged upside down between stems until it managed to wriggle free. They were noisy, graceless aeronauts, but functional. When they landed on a leaf, the leaf bowed or flipped with their weight. I saw more adults, and more and more, all ridiculously easy to grab and manipulate. The first one that I picked up didn't seem to want to let me go; it clung to my fingers with its spiky feet and made no move to fly away. After a few seconds, I realized that two of its four wings were stuck together. It was the first of many molting accidents that I would see.

Peril and abundance

On a walk around the neighborhood, I witnessed a staggering amount of cicada carnage. There were body fragments everywhere. A cicada still wiggling, with its head bitten off. Squished cicadas smothered in ants. A recently molted cicada, wings orange and furled, body white, head also missing. The head is the easiest and most nutritious cut of cicada, and predators can afford to be choosy with such an abundance on offer.

I passed a tree wrapped in netting to protect against egg laying. It would be days before most females were ready to lay, but the owners of this tree were preparing for the eventual influx of eggs that can damage and destroy whole branches. While dung beetles' parents dig into the ground, and wasp mothers inject eggs into caterpillars, cicada moms see homes for their babies in trees. Under the bark is a haven of safety, hidden from predators and kept moist by the living tissue of the tree. But accessing it presents a serious engineering trial. Female cicadas rise to the challenge with metal-reinforced ovipositors—structures for laying eggs. They're made of the same stuff, chitin, as all insect exoskeletons, but at the edges where the sawing action takes place, the chitin is impregnated with manganese and zinc.[3]

Millions of cicadas, each laying hundreds of eggs, can have a significant environmental impact. Those metallic ovipositors sometimes cut deeply enough into a tree that a whole branch browns and withers, a phenomenon called flagging. Humans historically assumed it was harmful, and research has shown that at least some species of tree do respond with defensive chemicals, but other trees have been reported to flourish after such natural pruning. Kritsky describes a tree in his neighborhood that received "intense damage" from cicada ovipositors during one emergence, then bloomed in abundance the following year. "It was like a snowball with a trunk."[4] Many humans have long been repulsed by cicadas. Even as an avowed invertebrate lover, I can admit that it was a bit startling to walk through a world so blatantly drenched in bugs. But my primary reaction as I passed the many bitten, squished, and otherwise mangled cicadas was sympathy.

"Hey buddy, why didn't you metamorphose?" I asked one of the dead nymphs littering the sidewalk. I saw another crawling along with its wings too wrinkled to fly. They hadn't expanded when they were supposed to, and now they never would. "Oh, your poor wings. Was it too cold for you last night? I was cold, until my friend brought me a hot water bottle."

I remembered how Williams described cicada metamorphosis to me on the phone. "Once that starts, it's like labor. There's no holding it back."[5] A cicada nymph, newly emerged from the ground,

must find a place to cling tightly as it goes through extreme con-
tortions. Its skin splits at the back, and the animal begins to work
its way out. Imagine that you're wearing a full-body jumpsuit that
unzips between your shoulders and sacrum, and you have to draw
your whole body out through that opening. Also, the jumpsuit covers your feet, hands, head, and face and is tightly molded to every
contour of your body, down to the eyeballs.

"For any arthropod, that's the most dangerous part of its life,"
says Williams. If part of the body gets stuck in the old skin, the
animal's a goner. It reminds me of the high hatching mortality of
birds, the chicks who aren't able to extricate themselves from their
shells and die within them. These tiny cicada tragedies were easy to
spot, because they couldn't fly away to hide in bushes and trees. But
as I wandered from block to block and peered more closely into the
foliage, I saw plenty of fresh adults that had molted to perfection.
Now they were still. Waiting. Quiet. Too soft inside to sing. Their
exoskeleton had hardened, but their internal skeleton needed to
stiffen up before they could make a proper racket. What was it like,
I wondered, to feel hard on the outside and soft on the inside?

In the evening, I called my family on video so they could see the
molts in the garden. As my children admired an empty skin, I spotted a nymph crawling right up out of the ground, and exclaimed
with excitement. I showed it to them, then hung up so I could take
pictures. As I squatted to watch one nymph trundle over the dirt,
I heard a rustle, looked over, and there was another. Some movement caught my eye. It was yet another. I realized that I sat in a
field of cicada nymphs, all lumbering along in search of a good spot
to pull free of their old skin. The ground everywhere was pocked
with holes, escape tunnels that the nymphs had been digging for
weeks, then waiting inside of for just the right time. I stuck my
pinky finger in an empty hole. The walls were hard, firm, and dry.

I stood there in the dark, walking from tree to bush to fence,
watching cicada after cicada struggle free from the last vestige of

its seventeen-year-long youth. When they do successfully molt, the transformation is stunning (see insert, photo 16). Wings that had been scrunched tightly inside the old skin are fully inflated and "come down like shower curtains," as Williams put it.[6]

As for the sound? The next day, it hit me as soon as I stepped outside. A cicada chorus. A high, persistent, and extremely localized hum. Sometimes it was background noise, but then I'd walk another block, find myself under trees thick with cicadas, and the noise would land squarely in the foreground. I could almost feel myself vibrating in tune with them. One longtime resident who'd worked in children's theater told me that during the last emergence they had to cancel a show because the performers couldn't hear the music and the audience couldn't hear the performers. He added, "At the peak of it, you'd come out and there would be cicadas on your tires. You'd have to brush them off—or just start driving."

When I went to visit the local entomologist Dan Gruner, I was touched to learn that no matter how abundant the cicadas, he always removes them from his tires before starting the car. In fact, when I arrived, he was busy picking up cicadas from the sidewalk and putting them on a tree. I commented on it appreciatively, and he answered in earnest, "I always save them. I went on a walk with some neighborhood kids this morning, six-year-olds. I was answering questions and showing them how to pick up the cicadas and put them on a tree to save them. Even though it makes absolutely no difference."[7] In this case, saving individual animals isn't about conserving the species, as it is for condors where every egg matters. It's about creating the kind of world you want to live in, Gruner and I agreed. A world where children choose to save cicadas rather than pull off their legs in stereotypical playground cruelty.

I ask Gruner the perennial question of how nymphs mark the passage of time. He suggests that the change in seasons might tick over a genetic marker. This could be another case where methylation of genes comes into play, but that remains an untested idea.

There are still so many mysteries attending cicada cycles. Gruner pointed out that everyone quotes 64°F (18°C) as the "magical, mystical" temperature that sparks emergence. But how do the cicadas know it's exactly 64°F? "They don't," Gruner told me cheerfully. That number was published in 1968, and the study has never been replicated. He and his students have been tracking soil temperatures assiduously, to gather data toward a more detailed view.

The connection between temperature and emergence isn't merely an academic subject. If nymphs are tightly dependent on temperature cues, then temperature differences between suburbs and forests may cause asynchrony: cicadas of the same brood emerging earlier in some places and later in others. This is a dangerous possibility for a species that depends on overwhelming predators with sheer numbers. If the emergence is too spread out, predators will never reach satiety.

To compare cicada habitats, we hopped a fence at the back of his property. On the other side, a trail led into a forested area, which I found noticeably cooler than the suburban yard and street had been. We saw many nymph holes along the trail, but few adults. Clearly, the emergence hadn't yet ramped up in here. Gruner took me to a tree surrounded by a thick layer of fallen leaves, and he showed me where to pull back the leaves to see chimneys or extensions of nymph tunnels. (Why only some tunnels get built up into chimneys is another mystery. One hypothesis is that it protects the nymphs from running water and rain.) I didn't see any nymphs inside chimneys or tunnels, but Gruner told me they hide deeper during the day. To avoid us lumbering numbskulls, no doubt.

Before I left, Gruner showed me a mourning dove nest in a planter hanging by his front door. He'd been carefully watering around the nest for weeks and was even able to snap a picture of the pale eggs inside when the brooding mother took a break to go foraging. Unlike many other birds, mourning doves aren't insect eaters, so she wouldn't be gorging herself on cicadas. She had something in common with the insects, though: the ability to adapt her reproductive habits to a human-dominated environment. She'd nested by a house, while the cicadas filled the tree-lined residential streets.

Buggy people

After visiting Gruner, I roped my friend Laura into driving out to Annapolis with me. I wanted to meet Williams in person, and I was particularly keen to witness the "ci-cow-das" she'd described to me on the phone: two large cow statues that flank the driveway to the Maryland Department of Agriculture, currently dressed up in cicada costumes to celebrate Brood X.

Williams embodies the dichotomy of a government entomologist in the time of emergence. She's simultaneously sick of talking about cicadas, and eager to talk your ear off about cicadas. Previous emergences wore her out with a constant stream of phone calls. Despite this, she told me, "I'm enthusiastic! I'm printing a few T-shirts, I've revised my cicada origami."[8] She's eager to search for the many color variants that can be spotted when you're surrounded by millions upon millions of cicadas. Instead of the usual red eyes, rare forms have blue or brown peepers. Williams fondly remembers one cicada she found in 2004: "Every vein in the wings was black. It was very hot, like a black Corvette." (It took me a second to realize she meant "hot" like *that*.)

Williams expresses concern about the future of the cicadas. "There's been an awful lot of developing since 2004, and you know the climate change thing is not exactly a rumor," she told me dryly. That said, different stages of human land development have different consequences. During early European settlement, forests were cut down and replaced with farmland, a massive loss of trees that did the cicadas no favors. These days, people are more likely to replace farmland with housing and shopping, which often includes planting ornamental or shade trees—a boon for cicadas. However, humans can't promise to leave a tree in place for seventeen years. If a suburban tree is cut down between emergences, any nymphs that had been suckling at its roots are almost certainly doomed.

In 2004, Williams found one particular tree at a private home that was so full of screaming cicadas she made a record volume measurement beneath it: 105 decibels, louder than a jackhammer, a volume that can damage human eardrums with prolonged exposure. As she described the experience, "It's painful. I have to wear ear plugs. At some point, my skin started to crawl, and I started feeling really queasy."[9] But as unpleasant as it surely is to be a human in the midst of such an insect din, I felt a different kind of queasy to learn the fate of that particular tree. The owner of the house passed away, the place was sold, and the new residents had the tree cut down so they could grow grass.

I became particularly grateful that Laura had agreed to come along when we hit traffic on the Beltway from DC out to Annapolis. Eventually our progress ground to a complete halt. As a Los Angelena, I have no problem with traffic, but as a clueless out-of-towner, I appreciated handing navigation over to a local. When Williams called to find out what was taking us so long, Laura answered the phone for me, patiently taking down the entomologist's directions, which were flavored with salty commentary about everything from the traffic to her dog's insatiable appetite for cicadas.

When we finally arrived, I was enchanted by the ci-cow-das. Large transparent wings were tied to their backs, black tights covered their noses, and red solo shot-glass cups bugged out over their eyes. Williams first made the costumes in 2004, and pointed out to us that, "Back then, they didn't have the small cups. So I used the big ones." Somehow, nothing says "seventeen years passed" like this simple statement.

As she guided us around, Williams kept us off the grass, which was apparently tick heaven. I found it very funny that the Department of Agriculture's lawn was full of ticks. While I was giggling,

Laura offered the practical suggestion that they get some opossums to eat the ticks. Williams countered that these ticks were too small to be slurped up by opossums. Larvae-obsessed as I am, I made a note to investigate the question later. Back home, I learned that the idea of opossums eating ticks originated with a 2009 study in which scientists dropped a bunch of larval ticks—which have only six legs, instead of the eight-legged adults—onto different mammals. After four days, they counted how many tick larvae remained in the enclosure with each animal, and concluded that any missing ticks had been eaten. This calculation gave opossums the impressive statistic of consuming about ninety percent of their tick larvae![10] Follow-up investigations, however, revealed that many ticks may have gone undetected. And when the stomach contents of wild opossums were examined, not a single tick adult or larva or even tick fragment could be found. Alas, opossums may not be tick-devouring superheroes after all.[11]

As we listened to the real cicadas chorus from our position beside the ci-cow-das, Williams observed with a straight face, "The hills are alive." But like me, she was disappointed that they weren't even louder. Several days later, after my return home, she emailed me triumphantly with the news that she'd finally gotten a decibel reading over 100. Since we had bonded over a mutual love of bugs and art, I responded by sending her the sketch of a cicada that she'd admired in my notebook. She, in turn, made me a cryptic offer: "If you go to Goodwill and find a light-color cotton T-shirt . . . and send me w/your address . . . something may happen."

One doesn't pass up such an opportunity. I mailed off a light gray shirt. It was returned to me with a beautifully screen-printed orange and black cicada above the words "The First Gen X."

Predators and parasites

With all the other omnivores in the area eagerly chowing down on cicadas, the question had to arise: what about humans? Insect protein, as we've seen, can be a more sustainable food than many other options, and despite an aversion to it in many modern industrialized societies, it has been a dietary staple in many parts of the world for a very long time.

Near the end of my time in DC, a friend alerted me to an imminent "pop-up cicada cookout." It was being organized by Bun Lai, a celebrity chef and leader in sustainable food, and it turned out to be located mere minutes from where I was staying. My lifelong vegetarianism briefly warred with my writer's instinct to immerse myself in the cicada experience as thoroughly as possible.

Time to eat some cicadas, I decided.

It was the last day of my trip, a spring scorcher. I was tempted to drive to the park with the AC on, but instead I rolled down the windows to savor the humid heat and the cicada vibrato. Driving from one tree-lined block to the next, I heard the sound swell and fade, like waves at sea.

At the park, I found a small and friendly group, including the friend who'd called me in, her children, some friends of Lai's, and various people who happened to be in the area. One young man, a camp counselor and gymnastics coach, remembered the 2004 emergence from his middle school years. He had no cicada-loving mentor like Gruner and wryly recounted, "We figured out that if you take off the front two legs, the head pops off—a fact I've been resisting communicating to the kids I teach." Instead, he chooses now to talk to them about empathy.

Even though we are there to eat cicadas, there is a gentleness, almost a reverence to the group. "This is a sacred experience," says Lai, though he grins as he says it, like it's a joke but maybe also the truth, inviting us all to decide for ourselves. "In Japan, people really embrace insects. We used to have them as pets. We had a rhinoceros beetle; it ate watermelon. I grew up loving insects, not really eating them."

Now, Lai sees insect protein as one of many approaches to sustainable eating. He often focuses on eating invasive species, which the cicadas are not. But with the cicada spectacle as a starting point, he can discuss opening our palates to many unconventional tastes in the name of a more environmentally compatible diet. Today, cicadas, tomorrow, perhaps black soldier fly larvae.

Lai had a small stove set up on a picnic table, and all the fixings to make sushi with the cicadas he fried. Following others before me, I laid seaweed on a plate, added rice, some avocado, then carefully selected two dead cicadas. When I ate, I pretended I was a bird. It tasted like avocado sushi (my favorite) with a little extra salt and crunch.

I soaked up the surreal, and also very real, experience of deliberately consuming animals for the first time in my life while surrounded by an abundance of said animals too large for me to fully grasp. I listened to them chorusing in the millions, perhaps the billions, though their density in the park itself was nothing overwhelming.

It wasn't until one last stop, on my way home from the cookout, that I found myself in such a thickness of cicadas that I simply couldn't avoid stepping on them, nor them on me. The location was the house of another friend—a developmental biologist, humorously enough. Two of her kids joined me in the backyard. I taught them how to identify male and female cicadas, and how to surprise males into squawking alarm calls, something that Williams had showed me. (There's nothing quite like holding a cicada up to your ear and feeling the gentle caress of its clawed feet against your cheek or catching in your hair as it squeaks about the indignity of the position.) In exchange, the kids helped me spot all the sights: pairs of mating cicadas joined rump-to-rump! A blue-eyed cicada! A cicada whose back half had been completely taken over by a crumbly yellow fungus!

Thoroughly satisfied with everything I'd managed to see and do, I flew home the next day, still pondering the periodical cicada phenomenon. Some species have an outsize impact on their ecosystems, like sunflower stars controlling sea urchin abundance. Cicadas probably aren't as critical. But how can we really rank ecological importance? Scientists are still finding new connections between different life stages of cicadas and the animals, plants, fungus, and microbes around them. Adult bodies compost and enrich the earth. The digging of the juveniles aerates the soil. The laying of eggs in branches revivifies older trees. Meanwhile, birds that usually eat caterpillars switch to cicadas. At first that sounds good for the caterpillars, but without birds keeping their numbers down, their abundance may invite more parasitoid wasps than usual to fill their bodies with eggs. Gruner and his students have been counting parasitized caterpillars to find out if this is the case.

Although he's intimately familiar with predation from his research, Gruner's gentle nature still grieved when an act of predation affected his avian housemate. A few days after my visit, he wrote to me, "Something got to the dove. The bird has evacuated and there were eggshells below. I'd grown rather attached to our friend and was rooting hard for her success, and I am now sad when I see just the stupid flowers." I got a little teary myself, hearing about it—perhaps it compounded with the sympathy I'd been feeling for all the cicada nymphs that don't make it, that get stuck in their molts or squashed or eaten.

Life always carries on. Later, when Gruner sent some research papers in the mail, he added on a Post-it, "P.S. A Carolina Wren took over the planter, there are now chicks. Not a total loss with departure of mourning dove."

Most of the animals in the world are embryos or larvae. And most of them will not survive to adulthood. This is the bittersweet reality of ecological developmental biology. I think about all the dead embryos inside bird eggs and all the marine larvae devoured by planktivorous creatures (including by other larvae). My visit to the cicada emergence highlighted the universality of mortality.

"They, like the comets, make but a short stay with us," wrote the scientist Benjamin Banneker in the year 1800, about adult cicadas. "If their lives are short, they are merry. They begin to sing or make a noise from first they come out of the earth till they die."[12]

The comet comparison is an apt one, since comets continue to exist, traveling through space, even when we cannot see them. Observing the bounty of adults gave me a new awe for the unseen young multitude underground, steadily counting their way through the next seventeen years without any guidance from their parents. After reveling in the bug bonanza, I returned home to my own offspring, bringing a renewed gratitude that I'm still around to watch them grow, and they're still around to do the growing.

EPILOGUE

Our Quiet Dependence on Babies

Your children are not your children.
They are the sons and daughters of Life's longing for itself.

—Kahlil Gibran, "On Children"[1]

When I told people I was writing a book about animal babies, they often assumed it was a children's book. While I have no intention of keeping this book away from interested children (including my own), you may have noticed that its length and vocabulary are more typical of books for adults. Even so, I have hardly been able to cover the topic exhaustively. Instead I have offered representative species and stories, aiming to present animal babies at their most vulnerable and at their most vital. My goal has been to open up a world both much larger and much smaller than you may have known before.

It never stops astonishing me to think that every animal on the planet reproduces by reducing itself to a single cell. Every body type, from conch to condor, finds a way to build itself anew—not from scratch, but from a combination of internal and external signals and relationships. Eggs and embryos are hidden in pocket environments all around the world, from spider eggs in house walls to mosquito eggs in park puddles to all the eggs dancing on ocean currents and even on the wind. The future of each species depends on enough of these fragile babies getting access to air, food, water, and symbiotic partners.

Each baby is a needle stitching through the fabric of the world, binding together the microscopic and the immense, the proximal and the distant, the future and the past.

Nature and nurture throughout our lives

"I need not labor the point at this late date that the characters of the individual are the product of both its genetic make-up and its environment,"[2] wrote Thomas Hunt Morgan, the embryologist who became a founder of genetics, in 1932. And here I am, ninety years later, laboring the point. That's not because science has made no progress—it is precisely because we have learned so much about both genetics and the environment that I have a point to labor.

The endless pendulum swing of popularity between scientific fields makes periodic reminders necessary. In the wake of Morgan's own evangelical enthusiasm for genetics, genetic determinism gained a fierce following, and environmental influences on development were all but forgotten. Our reverence for genes arose alongside and became entangled with our information technology. As computers made it possible first to sequence genomes at all, then to do so more and more rapidly, we reached for computer metaphors to describe the genomes themselves. DNA became a "code" that was "run" by the body's "machinery." The metaphor may have its uses, but the more we learn, the more inaccurate it appears to be.

An alternate metaphor suggested by Gilbert is, "to see each holobiotic organism as a performance. We inherit a *score* (the genome), a means of *interpreting* the score, and a means of *improvisation* should the score be incomplete. . . . Each performance will be different, even if the DNA is the same. Identical twins are different performers of the same score."[3]

A computer program can be "run and done," but a performance can last a lifetime. In the case of development, it does. "I'm forty-nine years old and I'm still developing as an individual," says

Tomberlin, the entomologist. "I love the idea that I'm not static, I'm always changing."[4]

Metamorphosis can occur in adult animals, changing one body form to another. Many species undergo sex changes as adults, driven by a combination of internal and external factors. Depending on reproductive strategy, it can be advantageous for either females or males to be the larger sex, so as an animal grows, it switches. Egg number is the limiting factor in clownfish reproduction, and bigger females can make more eggs. Thus the largest clownfish are females, and if a female dies, the largest male becomes female. (After the opening scene of *Finding Nemo*—spoiler alert!—Nemo's dad, Marlin, would have become Nemo's mom.) Meanwhile, blue-head wrasse reproduction is shaped by male-defended territory, and bigger males can defend more territory. So when a large male dies, a female turns male. Such changes involve significant modifications, not only exchanging sex organs but also altering body shape, coloration, and behavior (see insert, photo 17).

Adult reproductive changes are not limited to fish. Chilean oysters begin their sexual maturity as males, releasing sperm into the water. As they age, they metamorphose into females, which will collect sperm from younger males—an obligate May-to-December relationship. After using this sperm to fertilize her eggs, the female oyster will care for them in the folds of her gills, which are heavily modified into brood chambers. Although I found deep breathing difficult toward the end of my pregnancies, due to the overall compression of my internal organs, I am grateful to have had one specific organ for growing a baby rather than having to modify my lungs, stomach, or mouth for that purpose.

We humans also go through well-documented developmental changes in adulthood, like menopause: a significant change in reproductive hormones and cessation of the female reproductive cycle. In Western science and medicine, menopause has typically been lumped with aging, or senescence. However, the evolution of menopause has also been proposed as an adaptation to increase reproductive success. This may seem paradoxical, but recall the wasp mothers who adjust the sex ratio of their offspring based on how many grandchildren each sex is likely to give her. The goal of reproduction is not merely children, but *descendants*. The "grandmother hypothesis" suggests that human babies and children are more likely to survive with extra care from a female grandparent, which could have favored the evolution of humans who stop their own reproductive output to invest in their children's children. "Why menopause?" remains an open question, but the possibilities are intriguing.[5]

The hormonal changes of menopause have been connected with various health problems, leading it to be treated as a pathology, but what if we change our perspective to see it as a developmental process? Throughout life, we constantly remodel our skeleton, breaking down bone material down and building it up. In menopause, this balance shifts toward breaking down, producing osteoporosis. However, after certain bacteria were shown to reverse this bone loss in mice, a control study in humans shows that specific probiotics are indeed able to protect against postmenopausal bone loss.[6]

If we think of the probiotic as a medical treatment, then it's easy to see menopause as a disease. On the other hand, microbes are our constant companions and coworkers throughout early development. Why would later stages of development be any different? Thus, we could think of the probiotic as environmental input, sought out and incorporated by our bodies for this next stage of developmental unfoldment. We could even think of it as a partial metamorphosis, the composition of ourselves as a holobiont shifting from one part of our life cycle to the next. Considering the ongoing conversations with our environment that we conduct throughout our lives, there need never be a particular age at which we declare ourselves complete. Each of us is an orchestra, performing the score that lasts a lifetime.

Secrets of health and longevity from babies

I'm not going to argue that all human health problems are simply misunderstood development—but I *am* inclined to suggest that knowledge of development will help us understand and treat nearly all health problems. Cancer, for example, can affect humans at any age but becomes more common as we grow older. And yet in the pithy words of one *New York Times* headline, "A Tumor [Is] the Embryo's Evil Twin."[7]

The cellular process that creates and maintains a malignant tumor is almost identical to the process of embryonic development. A reliable oxygen supply is crucial to developing tissues, so embryos have multiple fail-safes to ensure that blood flows where it's needed. Tumors exhibit the same fail-safe mechanisms, rerouting blood flow to themselves when medical treatments attempt to suffocate them. That's frustrating for patients and doctors, but the similarity between embryos and tumors can also point the way to anticancer drugs. Thalidomide, which caused thousands of birth defects from severe to fatal in 1957 through 1961, messes up tumors as well as it messed up embryos.

As unpleasant as it is, I need to tell you how scientists create tumors in adult rodents: They graft an early rodent embryo onto the organ where they want the tumor. Somewhat more pleasant to think about is this: Scientists have also experimented with grafting actual tumors onto rodent embryos and found that these tumors don't become malignant. The embryo simply absorbs the tumor and goes on with normal development. Embryo and tumor speak the same language, and the embryo knows how to run the show.

When tumors arise in adult life, in a sense it's misplaced embryonic development. Genes that were useful when we were first building our bodies become pathological when they're turned back on at the wrong time. However, if tumors prove that it's possible to reactivate the bodybuilding know-how of an embryo, that leads to a logical next question. "Why can't we regenerate?" asked Kathy Foltz, my first developmental biology professor. "Our livers do a pretty good job, and if you're fairly young, the tips of your fingers do a pretty good job, but the rest of it is pretty tough. But if we can have cancer, why can't we have the regeneration aspect?"[8]

Many researchers hope that someday we can. A greater understanding of development will inform not only our treatment of disease, but our dreams of regeneration. Our bodies know how to build themselves once. Perhaps we can learn how to access that knowledge to build damaged parts again.

For that matter, why should we even age? Development in other animals shows that the aging clock can be stopped—for days, for a season, for years. The molecules that roundworms use to produce their suspended-animation dauer larvae are also linked to longevity. Scientists have played with these molecular connections to create extraordinarily long-lived worms. Many invertebrates, like Strathmann's sea snails and Hodin's sunflower stars, can draw out their life cycle indefinitely during certain critical times, often just before or just after metamorphosis. The incredible immortal jellyfish can actually reverse-metamorphose, moving either forward or backward in its life cycle depending on environmental conditions. In theory, a jelly could keep doing this forever, although scientists don't know what the record actually is. (These jellies are still susceptible to all the usual death hazards of living: predator, pathogen, parasite, pollution . . .)

It isn't only invertebrates that may hold developmental secrets to longevity. Greenland sharks live for hundreds of years without the rising risks of heart disease and cancer that aging humans face. They also take by far the longest time of any vertebrate to reach maturity: 150 years. Although their biology may hold valuable clues to potential human perpetuity, probably none of us wants to spend a century and a half as a child.

With all the tantalizing possibilities on display in the rest of the animal kingdom, we do have to wonder: Could humans ever become young again?

Would we *want* to?

Let's take care of babies just because

I don't know how well you remember being a kid, but it's kind of a raw deal. Young animals have fewer options when it comes to habitat and diet, being limited by their small size. At the same time, they're more vulnerable to predators, temperature extremes, and environmental toxins. What's more, adults aren't hesitant to manipulate kids, or straight-up eat them if the situation calls for it—as in the case of milkweed butterflies drinking caterpillar blood, or mouth-brooding fish swallowing their eggs.

To be fair, we humans do have a strong urge to protect rather than consume our children. But this desire is often couched in terms of protecting their potential or honoring what they will become as adults. John F. Kennedy famously stated, "Children are the world's most valuable resource and its best hope for the future." This is a beautiful sentiment, and I might just add that children are worth protecting not only because they are the future but simply because they are alive here and now.

What would such protection entail for the young of all animals? Perhaps the most obvious problem that stands out from my research on developmental biology is the modern version of *Silent Spring*—the environmental impacts of endocrine-disrupting chemicals. There are so many of these chemicals and they are so widespread that the situation can seem overwhelming, but we don't have to know exactly which ones cause which developmental problems in which species in order to make changes for the better. To start, we could take greater control over agricultural runoff. Atrazine, the most widely used weed killer in the world, is already banned in Europe (including Switzerland, where it is also, ironically, manufactured). What if we phased it out in the rest of the world? And what if we *didn't* replace it with newer, not-yet-studied herbicides?

As a result of the furor over BPA, many products now advertised as "BPA-free" use chemicals that are probably just as disruptive but haven't been examined yet. Europe has a "dangerous until proven safe" policy for new chemicals, while the United States' policy is, regrettably, the other way around. "As if in a homicide case, the product is assumed to be innocent until proven guilty beyond a reasonable doubt," Gilbert writes dryly. Such proof is exceptionally difficult to acquire when wealthy corporations have a vested interest in the *absence* of proof. Although the power of the chemical industry can seem unstoppable, Gilbert goes on to suggest that individual people seemingly at odds with each other could band together to fight back. "The antiabortion bishop and the physician who prescribes contraceptives both claim to want healthy babies in healthy families. To this end, both can promote and support legislation to ban endocrine-disrupting chemicals."[9]

Human developmental processes are not distinct from those of the rest of the animal kingdom. When we act to create a safer environment for ourselves and our children, we will also be acting to protect the myriad species with which we share the planet. Those of us who experience no metamorphosis, who keep the same limbs and organs from womb to grave, are the exception rather than the rule. Still, our butt-first embryos link us to the rest of the animal kingdom, and we all face the same developmental challenges. As parents, how can we provide for our children while also preparing them for an independent life? As children, how can we make the best of what we're given and find our own place in the world?

Thinking of his work on dung beetle niche construction, Armin Moczek muses, "Probably all parents worry about this—you worry about your failure to provide the things they need. You wonder, will they ever learn this, or how will they navigate that. My daughter is now nineteen, my son is sixteen, and I've just been impressed by how gradually but steadily, and somehow unavoidably, they've become agents of their own environments. Who they want to be, what they want to be. That has put me at ease."[10] With my own children growing up ten years behind his, I engrave these words on my heart.

"I see the untarnished optimism of a young person and I love it and I try to encourage it," says Tomberlin. "That does give me hope for the human species. I love being a parent."[11]

I've found in raising children that one of my most valuable lessons came from an unorthodox source—acting. As a kid, I attended theater classes and summer Shakespeare camps. We often practiced our acting skills with improvisation exercises, where the fundamental rule is *yes, and*. In an improv game, you never contradict or refuse to play along with a scene partner. You take what they give you and build on it. This has informed my interactions with children, both my own and others, so I don't find myself denying or minimizing their needs.

"Yes, let's play pretend, and also let's head for the car because it's time to go home. I'll be the train engine! Want a ticket to ride?" I try to reinforce it in their interactions with their siblings and friends. "He says he really wants to be a cat. Can you think of a way to include that in your dragon game?"

Child's play? Yes. *And*, a way of doing science, a way of building a better world. "Yes, and" helps us resist our mental tendencies toward oversimplification and rigidity. Yes, genes are important, and so is the environment. Yes, organisms compete with each other, and they also cooperate. The more we practice this, I think, the richer and closer to truth our view of the world becomes.

NOTES

INTRODUCTION | A World of the Babies, by the Babies, for the Babies

1. W. Garstang et al., *Larval Forms, and Other Zoological Verses* (Chicago: University of Chicago Press, 1985).

2. S. Gilbert, interview with the author via Zoom, November 18, 2021.

3. J. M. Biesterfeldt et al., "Prevalence of Chemical Interference Competition in Natural Populations of Wood Frogs, Rana Sylvatica," *Copeia* 3 (1993): 688–95.

4. M. F. Benard, "Warmer Winters Reduce Frog Fecundity and Shift Breeding Phenology, Which Consequently Alters Larval Development and Metamorphic Timing." *Global Change Biology* 21, no. 3 (2015): 1058–65.

5. D. J. Messmer et al., "Plasticity in Timing of Avian Breeding in Response to Spring Temperature Differs Between Early and Late Nesting Species," *Scientific Reports* 11, no. 1 (2021): 5410.

6. "The Salmon Life Cycle—Olympic National Park," National Park Service, accessed March 22, 2022, nps.gov/olym/learn/nature/the-salmon-life-cycle.htm.

7. A. Purser et al., "Vast Icefish Breeding Colony Discovered in the Antarctic," *Current Biology* 32, no. 4 (2022): 842–50.

8. "POP2 Children as a percentage of the population," Childstats, accessed March 22, 2022, childstats.gov/americaschildren/tables/pop2.asp.

9. "Average age by country," WorldData.info, accessed March 3, 2022, worlddata. info/average-age.php.

10. S. F. Gilbert and D. Epel, *Ecological Developmental Biology: The Environmental Regulation of Development, Health, and Evolution*, Second Edition (New York: Oxford University Press, 2015).

11. Ibid.

12. Ibid.

13. "Temperature-Dependent Sex Determination: Current Practices Threaten Conservation of Sea Turtles," Science, accessed December 3, 2021, science.org/doi/abs/10.1126/science.7079758.

14. R. A. Relyea, "Predator Cues and Pesticides: A Double Dose of Danger for Amphibians," *Ecological Applications* 13, no. 6 (2003): 1515–21.

15. J. M. Kiesecker, "Synergism Between Trematode Infection and Pesticide Exposure: A Link to Amphibian Limb Deformities in Nature?," *Proceedings of the National Academy of Sciences* 99, no. 15 (2002): 9900–4.

16. Gilbert, 2015

17. "Fisheries & Aquaculture—Fishery Statistical Collections fact sheets—Global

Capture Production," Food and Agriculture Organization of the United Nations, accessed December 3, 2021, fao.org/fishery/en/collection/capture.

18. D. J. Staaf et al., "Natural Egg Mass Deposition by the Humboldt Squid (*Dosidicus gigas*) in the Gulf of California and Characteristics of Hatchlings and Paralarvae," *Journal of the Marine Biological Association of the United Kingdom* 88, no. 4 (2008): 759–70.

19. T. Vendl, and P. Šípek, "Immature Stages of Giants: Morphology and Growth Characteristics of Goliathus Lamarck, 1801 Larvae Indicate a Predatory Way of Life (Coleoptera, Scarabaeidae, Cetoniinae)," *Zookeys* 619 (2016): 25–44.

20. S. B. Emerson, "The Giant Tadpole of *Pseudis Paradoxa*," *Biological Journal of the Linnean Society* 34, no. 2 (1988): 93–104.

21. P. M. Stepanian et al., "Declines in an Abundant Aquatic Insect, the Burrowing Mayfly, across Major North American Waterways," *PNAS* 117, no. 6 (2020): 2987–92.

22. A. Lovas-Kiss et al., "Experimental Evidence of Dispersal of Invasive Cyprinid Eggs inside Migratory Waterfowl," *PNAS* 117, no. 27 (2020): 15397–99.

23. R. Collin et al., "World Travelers: DNA Barcoding Unmasks the Origin of Cloning Asteroid Larvae from the Caribbean," *The Biological Bulletin* 239, no. 2 (2020): 73–79.

24. D. M. Ripley et al., "Ocean Warming Impairs the Predator Avoidance Behaviour of Elasmobranch Embryos," *Conservation Physiology* 9, no. 1 (2021).

25. J. L. Savage et al., "Low Hatching Success in the Critically Endangered Kākāpō (*Strigops habroptilus*) Is Driven by Early Embryo Mortality Not Infertility," bioRxiv (2020).

26. L. Yang et al., "Biodegradation of Expanded Polystyrene and Low-Density Polyethylene Foams in Larvae of Tenebrio Molitor Linnaeus (Coleoptera: Tenebrionidae): Broad versus Limited Extent Depolymerization and Microbe-Dependence versus Independence," *Chemosphere* 262 (2021): 127818.

27. S. J. Song et al., "Naturalization of the Microbiota Developmental Trajectory of Cesarean-Born Neonates after Vaginal Seeding," *Med* 2, no. 8 (2021): 951–64.

CHAPTER 1 | Eggs: Not Just a Bird Thing

1. Cadamole, "A Biologist's Mother's Day Song," last modified May 8, 2010, youtube. com/watch?v=osWuWjbeO-Y.

2. R. H. Harris and M. Emberley, *It's So Amazing!: A Book about Eggs, Sperm, Birth, Babies, and Families* (Somerville, MA: Candlewick Press, 1999).

3. "Leading egg producing countries worldwide, 2020," Statista, accessed May 18, 2022, statista.com/statistics/263971/top-10-countries-worldwide-in-egg-production.

4. P. Sutovsky et al., "Ubiquitin Tag for Sperm Mitochondria," *Nature* 402, no. 6760 (1999): 371–72.

5. S. Breton et al., "The Unusual System of Doubly Uniparental Inheritance of MtDNA: Isn't One Enough?" *Trends in Genetics* 23, no. 9 (2007): 465–74.

6. W. Fan et al., "A Mouse Model of Mitochondrial Disease Reveals Germline Selection Against Severe MtDNA Mutations," *Science* 319, no. 5865 (2008): 958–62.

7. Cadamole, 2010.

8. O. A. Ryder et al., "Facultative Parthenogenesis in California Condors," *Journal of Heredity* 112, no. 7 (2021): 569–74.

9. A. Soubry, "Epigenetic Inheritance and Evolution: A Paternal Perspective on Dietary Influences," *Progress in Biophysics and Molecular Biology* 118, no. 1 (2015): 79–85.

10. M. A. Birk et al., "Observations of Multiple Pelagic Egg Masses from Small-Sized Jumbo Squid (*Dosidicus gigas*) in the Gulf of California," *Journal of Natural History* 51, nos. 43–44 (2017): 2569–84.

11. E. E. Just, *The Biology of the Cell Surface* (Philadelphia: The Technical Press, 1939).

12. E. E. Just, "Cortical Cytoplasm and Evolution," *The American Naturalist* 67, no. 708 (1933): 20–29.

13. S. Gilbert, interview with the author via Zoom, November 18, 2021.

14. Ibid.

15. S. F. Gilbert and D. Epel, *Ecological Developmental Biology: The Environmental Regulation of Development, Health, and Evolution*, Second Edition (New York: Oxford University Press, 2015).

16. I. Blickstein and L. G. Keith, "On the Possible Cause of Monozygotic Twinning: Lessons from the 9-Banded Armadillo and from Assisted Reproduction," *Twin Research and Human Genetics* 10, no. 2 (2007): 394–99.

17. S. Sumner et al., "Why We Love Bees and Hate Wasps," *Ecological Entomology* 43, no. 6 (2018): 836–45.

18. A. A. Forbes et al., "Quantifying the Unquantifiable: Why Hymenoptera, Not Coleoptera, Is the Most Speciose Animal Order," *BMC Ecology* 18, no. 1 (2018): 21.

19. M. S. Smith et al., "*Copidosoma floridanum* (Hymenoptera: Encyrtidae) Rapidly Alters Production of Soldier Embryos in Response to Competition," *Annals of the Entomological Society of America* 110, no. 5 (2017): 501–5.

20. S. H. Orzack and E. D. Parker Jr., "Sex-Ratio Control in a Parasitic Wasp, Nasonia Vitripennis. I. Genetic Variation in Facultative Sex-Ratio Adjustment," *Evolution* 40, no. 2 (1986): 331–40.

21. D. Millerand and S. Adamo, "Parasitic wasps turn other insects into 'zombies,' saving millions of humans along the way," The Conversation, accessed December 8, 2021, theconversation.com/parasitic-wasps-turn-other-insects-into-zombies-saving-millions-of-humans-along-the-way-170610.

22. S. F. Gilbert, *Developmental Biology*, 6th ed. (Sunderland, MA: Sinauer Associates, 2000).

23. Ibid.

24. R. Grosberg, interview with the author via Zoom, August 18, 2021.

25. F. Oyarzun, 2021, interview with the author via Zoom, March 15, 2021.

26. E. Metchnikoff. "Untersuchung Uber Die Mesodermalen Phagocyten Einiger Wirbel Tiere," *Biologisches Centralblatt* 3 (1883): 560–65.

27. M. Poláčik et al., "Embryo Ecology: Developmental Synchrony and Asynchrony in the Embryonic Development of Wild Annual Fish Populations," *Ecology and Evolution* 11, no. 9 (2021): 4945–56.

28. N. Hemmings, 2021, interview with the author via Zoom, October 6, 2021.

29. Ibid.

30. J. L. Savage et al., "Low Hatching Success in the Critically Endangered Kākāpō (*Strigops habroptilus*) Is Driven by Early Embryo Mortality Not Infertility," bioRxiv (2020).

31. Hemmings, 2021.
32. K. M. Warkentin, "How Do Embryos Assess Risk? Vibrational Cues in Predator-Induced Hatching of Red-Eyed Treefrogs," *Animal Behaviour* 70, no. 1 (2005): 59–71.
33. R. M. Kempster et al., "Survival of the stillest: predator avoidance in shark embryos," *PLOS ONE* 8, no. 1 (2013): e52551.

CHAPTER 2 | Provisioning: From Edible Siblings to Algal Life-Support

1. R. Hayden, *A Ballad of Remembrance* (London: Paul Breman, 1962).
2. E. J. A. Cunningham and A. F. Russell, "Egg Investment Is Influenced by Male Attractiveness in the Mallard," *Nature* 404, no. 6773 (2000): 74–77.
3. A. Velando et al., "Pigment-Based Skin Colour in the Blue-Footed Booby: An Honest Signal of Current Condition Used by Females to Adjust Reproductive Investment," *Oecologia* 149, no. 3 (2006): 535–42.
4. R. Collin, interview with the author via Zoom, March 29, 2021.
5. R. Grosberg, interview with the author via Zoom, August 18, 2021.
6. F. X. Oyarzun and R. R. Strathmann, "Plasticity of Hatching and the Duration of Planktonic Development in Marine Invertebrates," *Integrative and Comparative Biology* 51, no. 1 (2011): 81–90.
7. F. X. Oyarzun, interview with the author via Zoom, March 15, 2021.
8. Grosberg, 2021.
9. Ibid.
10. C.-Y. Cai et al., "Early Origin of Parental Care in Mesozoic Carrion Beetles," *PNAS* 111, no. 39 (2014): 14170–74.
11. H. Vogel et al., "The Digestive and Defensive Basis of Carcass Utilization by the Burying Beetle and Its Microbiota," *Nature Communications* 8, no. 1 (2017): 15186.
12. A. P. Moczek, interview with the author via Zoom, September 24, 2021.
13. C. C. Ledón-Rettig et al., "*Diplogastrellus* Nematodes Are Sexually Transmitted Mutualists That Alter the Bacterial and Fungal Communities of Their Beetle Host, *PNAS* 115, no. 42 (2018): 10696–10701.
14. Moczek, 2021.
15. G. J. Dury et al., "Maternal and Larval Niche Construction Interact to Shape Development, Survival, and Population Divergence in the Dung Beetle *Onthophagus taurus*," *Evolution & Development* 22, no. 5 (2020): 358–69.
16. E. Snell-Rood, interview with the author via Zoom, August 19, 2021.
17. R. Osawa et al., "Microbiological Studies of the Intestinal Microflora of the Koala, Phascolarctos-Cinereus .2. Pap, a Special Maternal Feces Consumed by Juvenile Koalas," *Australian Journal of Zoology* 41, no. 6 (1993): 611.
18. F. Landmann et al., "Co-Evolution Between an Endosymbiont and Its Nematode Host: Wolbachia Asymmetric Posterior Localization and AP Polarity Establishment," *PLOS Neglected Tropical Diseases* 8, no. 8 (2014): e3096.
19. M. S. Gil-Turnes et al., "Symbiotic Marine Bacteria Chemically Defend Crustacean Embryos from a Pathogenic Fungus," *Science* 2246 (1989): 116–18.
20. T. J. Little et al., "Male Three-Spined Sticklebacks *Gasterosteus Aculeatus* Make Antibiotic Nests: A Novel Form of Parental Protection?," *Journal of Fish Biology* 73, no. 10 (2008): 2380–89.

21. K. Foltz, interview with the author via Zoom, March 29, 2021.

22. G. D. D. Hurst et al., "Male-Killing *Wolbachia* in Two Species of Insect," *Proceedings of the Royal Society B* 266, no. 1420 (1999): 735.

23. S. Leclercq et al., "Birth of a W Sex Chromosome by Horizontal Transfer of *Wolbachia* Bacterial Symbiont Genome," *PNAS* 113, no. 52 (2016): 15036–41.

24. T. B. Hayes et al., "Hermaphroditic, Demasculinized Frogs after Exposure to the Herbicide Atrazine at Low Ecologically Relevant Doses," *PNAS* 99, no. 8 (2002): 5476–80.

25. P. Salinas-de-León et al., "Deep-Sea Hydrothermal Vents as Natural Egg-Case Incubators at the Galapagos Rift," *Scientific Reports* 8, no. 1 (2018): 1788.

26. A. M. Hartwell et al., "Clusters of Deep-Sea Egg-Brooding Octopods Associated with Warm Fluid Discharge: An Ill-Fated Fragment of a Larger, Discrete Population?," *Deep Sea Research Part I: Oceanographic Research Papers* 135 (2018): 1–8.

27. S. Baillon et al., "Deep Cold-Water Corals as Nurseries for Fish Larvae," *Frontiers in Ecology and the Environment* 10, no. 7 (2012): 351–56.

28. N. Roux et al., "Sea Anemone and Clownfish Microbiota Diversity and Variation During the Initial Steps of Symbiosis," *Scientific Reports* 9, no. 1 (2019): 19491.

29. L. A. Levin and G. W. Rouse, "Giant Protists (Xenophyophores) Function as Fish Nurseries," *Ecology* 101, no. 4 (2020): e02933.

30. L. S. Zacher and R. R. Strathmann, "A Field Experiment Demonstrating Risk on the Seafloor for Planktonic Embryos," *Limnology and Oceanography* 63, no. 6 (2018): 2708–16.

31. A. Moran, interview with the author via Zoom, April 23, 2021.

32. Ibid.

33. Ibid.

CHAPTER 3 | Brooding Eggs: Carry Them, Sit on Them, Swallow Them Whole

1. R. Dove, *Mother Love: Poems* (New York: Norton, 1996).

2. P. Laval, "The Barrel of the Pelagic Amphipod *Phronima Sedentaria* (Forsk.) (Crustacea: hyperiidea)," *Journal of Experimental Marine Biology and Ecology* 33, no. 3 (1978): 187–211.

3. S. G. Nelson and C. O. Krekorian, "The Dynamics of Parental Care of *Copeina Arnoldi* (Pisces, Characidae)," *Behavioral Biology* 17, no. 4 (1976): 507–18.

4. J. Delia et al., "Patterns of Parental Care in Neotropical Glassfrogs: Fieldwork Alters Hypotheses of Sex-Role Evolution," *Journal of Evolutionary Biology* 30, no. 5 (2017): 898–914.

5. "'Rescuing' Baby Hummingbirds," Life, Birds, and Everything, last modified May 27, 2009, fieldguidetohummingbirds.wordpress.com/2009/05/27/rescuing-baby-hummingbirds.

6. "Black Phoebe," All About Birds, The Cornell Lab, accessed December 14, 2021, allaboutbirds.org/guide/Black_Phoebe/overview.

7. "Understanding an Ecological Trap," NestWatch, accessed November 23, 2021, nestwatch.org/connect/blog/understanding-ecological-trap.

8. W.-S. Huang and D. A. Pike, "Climate Change Impacts on Fitness Depend on Nesting Habitat in Lizards," *Functional Ecology* 25, no. 5 (2011): 1125–36.

9. A. Petherick, "A Solar Salamander," *Nature* (2010).

10. R. Kerney, "Intracellular Green Algae (*Oophilia amblystomatis*) in a Salamander Host (*Ambystoma maculatum*)," *The FASEB Journal* 25, no. S1 (2011): 420.2.

11. R. Collin, interview with the author via Zoom, March 29, 2021.

12. B. Robison et al., "Deep-Sea Octopus (*Graneledone boreopacifica*) Conducts the Longest-Known Egg-Brooding Period of Any Animal," *PLOS ONE* 9, no. 7 (2014): e103437.

13. M. W. Gray et al., "Life History Traits Conferring Larval Resistance Against Ocean Acidification: The Case of Brooding Oysters of the Genus Ostrea," *Journal of Shellfish Research* 28, no. 3 (2019): 751–61.

14. R. Strathmann, interview with the author via Zoom, March 5, 2021.

15. M. Fernández, interview with the author via Zoom, May 5, 2021.

16. Ibid.

17. Z. P. Burris, "Costs of Exclusive Male Parental Care in the Sea Spider *Achelia simplissima* (Arthropoda: Pycnogonida)," *Marine Biology* 158, no. 2 (2011): 381–90.

18. "The Long, Involved Process of Giant Water Bug Mating," The Dragonfly Woman, last modified November 7, 2011, thedragonflywoman.com.

19. C. R. Largiadèr et al., "Genetic Analysis of Sneaking and Egg-Thievery in a Natural Population of the Three-Spined Stickleback (*Gasterosteus aculeatus* L.)," *Heredity* 86, no. 4 (2001): 459–68.

20. R. Gloag, interview with the author via Zoom, August 26, 2021.

21. "California's Invaders: Brown-Headed Cowbird," accessed December 15, 2021, wildlife.ca.gov/Conservation/Invasives/Species/Cowbird.

22. Gloag, 2021.

23. Ibid.

24. Ibid.

25. Ibid.

26. Ibid.

27. A. P. Moczek, interview with the author via Zoom, September 24, 2021.

CHAPTER 4 | Pregnancy: Not Just a Mammal Thing

1. W. Whitman, from *Leaves of Grass*, bartleby.com/142/103.html.

2. D. G. Blackburn, "Evolution of Vertebrate Viviparity and Specializations for Fetal Nutrition: A Quantitative and Qualitative Analysis: Viviparity and Fetal Nutrition," *Journal of Morphology* 276, no. 8 (2015): 961–90.

3. A. N. Ostrovsky et al., "Matrotrophy and Placentation in Invertebrates: A New Paradigm," *Biological Reviews* 91, no. 3 (2016): 673–711.

4. "Prolonged Milk Provisioning in a Jumping Spider," *Science* 362, no. 6418 (2018): 1052–55.

5. W. Osterloff, "Do Sharks Lay Eggs?," Natural History Museum, accessed December 12, 2021, nhm.ac.uk/discover/do-sharks-lay-eggs .html.

6. K. Sato et al., "How Great White Sharks Nourish Their Embryos to a Large Size: Evidence of Lipid Histotrophy in Lamnoid Shark Reproduction," *Biology Open* 5, no. 9 (2016): 1211–15.

7. D. G. Swift et al., "Evidence of Positive Selection Associated with Placental Loss in Tiger Sharks," *BMC Evolutionary Biology* 16 (2016): 126.

8. P. R. Bell et al., "Oldest preserved umbilical scar reveals dinosaurs had 'belly buttons,'" *BMC Biology* 20, no. 1 (2022): 1–7.

9. Blackburn, "Evolution of Vertebrate Viviparity."

10. G. Quirós, "A Tsetse Fly Births One Enormous Milk-Fed Baby," last modified January 28, 2020, KQED, kqed.org/science/1956004/a-tsetse-fly-births-one-enormous-milk-fed-baby.

11. "Bursting with Babies: Bizarre Reproduction Contributes to Mite's Rapid Population Growth," *Entomology Today*, accessed June 1, 2022, entomologytoday.org/2017/05/16/bursting-with-babies-bizarre-reproduction-contributes-to-mites-rapid-population-growth.

12. S. Gilbert, interview with the author via Zoom, November 18, 2021.

13. Ibid.

14. Ibid.

15. A. R. Chavan et al., "Evolution of Embryo Implantation Was Enabled by the Origin of Decidual Stromal Cells in Eutherian Mammals," *Molecular Biology and Evolution* 38, no. 3 (2021): 1060–74.

16. P. A. Nepomnaschy et al., "Cortisol Levels and Very Early Pregnancy Loss in Humans," *PNAS* 103, no. 10 (2006): 3938–42.

17. S. A. Wahaj et al., "Siblicide in the Spotted Hyena: Analysis with Ultrasonic Examination of Wild and Captive Individuals," *Behavioral Ecology* 18, no. 6 (2007): 974–84.

18. H. J. Blom et al., "Neural Tube Defects and Folate: Case Far from Closed." *Nature Reviews Neuroscience* 7, no. 9 (2006): 724–31.

19. R. Morello-Frosch et al., "Environmental Chemicals in an Urban Population of Pregnant Women and Their Newborns from San Francisco," *Environmental Science & Technology* 50, no. 22 (2016): 12464–72.

20. S. F. Gilbert and D. Epel, *Ecological Developmental Biology: The Environmental Regulation of Development, Health, and Evolution*, Second Edition (New York: Oxford University Press, 2015).

21. R. B. Lathi et al., "Conjugated Bisphenol A in Maternal Serum in Relation to Miscarriage Risk," *Fertility and Sterility* 102, no. 1 (2014): 123–28.

22. Gilbert, *Ecological Developmental Biology*.

23. A. Hamdoun, interview with the author via Zoom, July 22, 2021.

24. "What is CRISPR," New Scientist, accessed January 27, 2022, newscientist.com/definition/what-is-crispr.

25. Hamdoun, 2021.

26. Ibid.

27. T. S. Stappenbeck et al., "Developmental Regulation of Intestinal Angiogenesis by Indigenous Microbes via Paneth Cells," *PNAS* 99, no. 24 (2002): 15451–5.

28. S. Gilbert, interview with the author via Zoom, November 18, 2021.

29. K. Korpela et al., "Maternal Fecal Microbiota Transplantation in Cesarean-Born Infants Rapidly Restores Normal Gut Microbial Development: A Proof-of-Concept Study," *Cell* 183, no. 2 (2020): 324–34.

30. C. Bondar, *Wild Moms* (New York: Simon and Schuster, 2018).

31. A. Ardeshir et al., "Breast-Fed and Bottle-Fed Infant Rhesus Macaques Develop Distinct Gut Microbiotas and Immune Systems," *Science Translational Medicine* 6, no. 252 (2014): 252ra120.

32. A. Kupfer et al., "Parental Investment by Skin Feeding in a Caecilian Amphibian," *Nature* 440, no. 7086 (2006): 926–29.

33. R. Emlet, interview with the author via Zoom, April 2, 2021.

34. S. Suzuki et al., "Matriphagy in the Hump Earwig, *Anechura harmandi* (Dermaptera: Forficulidae), Increases the Survival Rates of the Offspring," *Journal of Ethology* 233, no. 2 (2005): 211–13.

CHAPTER 5 | Unaccompanied Minors: Where Do the Escargot?

1. K. Gibran, "The Farewell," poets.org/poem/farewell-2.

2. C. R. O'Neill and A. Dextrase, "The Introduction and Spread of the Zebra Mussel in North America," Proceedings of the Fourth International Zebra Mussel Conference, Madison, Wisconsin, March 1994: 14.

3. R. Emlet, interview with the author via Zoom, April 2, 2021.

4. F. S. Chia and J. G. Spaulding, "Development and Juvenile Growth of the Sea Anemone, *Tealia crassicornis*," *The Biological Bulletin* 142, no. 2 (1972): 206–18.

5. "Gypsy Moths," Smithsonian Institution, accessed June 1, 2022, si.edu/spotlight/buginfo/gypsy-moths.

6. H. Fuchs, interview with the author via Zoom, October 20, 2021.

7. L. Winhold, "Unionidae," Animal Diversity Web, accessed June 1, 2022, animaldiversity.org/accounts/Unionidae.

8. A. Lovas-Kiss et al., "Experimental Evidence of Dispersal of Invasive Cyprinid Eggs inside Migratory Waterfowl," *PNAS* 117, no. 27 (2020): 15397–99.

9. "Capitula on Stick Insect Eggs and Elaiosomes on Seeds: Convergent Adaptations for Burial by Ants," *Functional Ecology* 6, no. 6 (1992): 642–48.

10. L. S. Mullineaux, interview with the author via Zoom, September 3, 2021.

11. T. R. Anderson and T. Rice, "Deserts on the Sea Floor: Edward Forbes and His Azoic Hypothesis for a Lifeless Deep Ocean," *Endeavour* 30, no. 4 (2006): 131–37.

12. F. Pradillon et al., "Developmental Arrest in Vent Worm Embryos, *Nature* 413, no. 6857 (2001): 698–99.

13. C. M. Young et al., "Embryology of Vestimentiferan Tube Worms from Deep-Sea Methane/Sulphide Seeps," *Nature* 381, no. 6582 (1996): 514–16.

14. A. G. Marsh et al., "Larval Dispersal Potential of the Tubeworm Riftia Pachyptila at Deep-Sea Hydrothermal Vents," *Nature* 411, no. 6833 (2001): 77–80.

15. Fuchs, 2021.

16. R. Collin, interview with the author via Zoom, March 29, 2021.

17. D. Vaughn, D. and R. R. Strathmann, "Predators Induce Cloning in Echinoderm Larvae," *Science* 319, no. 5869 (2008): 1503.

18. K. A. McDonald and D. Vaughn, "Abrupt Change in Food Environment Induces Cloning in Plutei of Dendraster Excentricus," *The Biological Bulletin* 219, no. 1 (2010): 38–49.

19. R. Collin et al., "World Travelers: DNA Barcoding Unmasks the Origin of Cloning Asteroid Larvae from the Caribbean," *The Biological Bulletin* 239, no. 2 (2020): 73–79.

20. A. Hamdoun, interview with the author via Zoom, July 22, 2021.

21. A. Moran, interview with the author via Zoom, April 23, 2021.

22. Ibid.

23. W. Garstang et al., *Larval Forms, and Other Zoological Verses* (Chicago: University of Chicago Press, 1985).

24. D. Zacherl et al., "The Limits to Biogeographical Distributions: Insights from the Northward Range Extension of the Marine Snail, *Kelletia kelletii* (Forbes, 1852)," *Journal of Biogeography* 30, no. 6 (2003): 913–24.

25. M. Álvarez-Noriega, interview with the author via Zoom, April 15, 2021.

26. H. L. Fuchs et al., "Wrong-Way Migrations of Benthic Species Driven by Ocean Warming and Larval Transport," *Nature Climate Change* 10, no. 11 (2020): 1052–56.

27. Collin, 2021.

CHAPTER 6 | It's Just a Phase: Why Babies Look Like Aliens

1. R. Graves, "The Caterpillar," Academy of American Poets, accessed June 1, 2022, poets.org/poem/caterpillar.

2. D. J. Staaf et al., "Natural Egg Mass Deposition by the Humboldt Squid (*Dosidicus gigas*) in the Gulf of California and Characteristics of Hatchlings and Paralarvae," *Journal of the Marine Biological Association of the United Kingdom* 88, no. 4 (2008): 759–70.

3. D. J. Marshall et al., "Developmental Cost Theory Predicts Thermal Environment and Vulnerability to Global Warming," *Nature Ecology and Evolution* 4, no. 3 (2020): 406–11.

4. D. J. Marshall, interview with the author via Zoom, September 23, 2021.

5. C. Goldberg, "Scientist at Work: Anne Simon; The Science Adviser to Whaaat?" *The New York Times*, January 6, 1998.

6. R. Emlet, interview with the author via Zoom, April 2, 2021.

7. F. X. Oyarzun, interview with the author via Zoom, March 15, 2021.

8. R. Strathmann, interview with the author via Zoom, March 5, 2021.

9. H. Karp, *The Happiest Baby on the Block: The New Way to Calm Crying and Help Your Newborn Baby Sleep Longer* (New York: Bantam Books, 2015).

10. D. J. Marshall, interview with the author via Zoom, September 23, 2021.

11. S. F. Gilbert and D. Epel, *Ecological Developmental Biology: The Environmental Regulation of Development, Health, and Evolution*, Second Edition (New York: Oxford University Press, 2015).

12. Ibid.

13. A. P. Moczek, interview with the author via Zoom, September 24, 2021.

14. S. Gilbert, interview with the author via Zoom, November 18, 2021.

15. M. Byrne, interview with the author via Zoom, April 19, 2021.

16. T. J. Carrier et al., "Microbiome Reduction and Endosymbiont Gain from a Switch in Sea Urchin Life History," *PNAS* 118, no. 16 (2021): e2022023118.

17. J. W. Brandt et al., "Culture of an Aphid Heritable Symbiont Demonstrates Its Direct Role in Defence against Parasitoids," *Proceedings of the Royal Society B: Biological Sciences* 284, no. 1866 (2017): 1925.

18. Gilbert, 2021.

19. G. Sharon et al., "Commensal Bacteria Play a Role in Mating Preference of *Drosophila melanogaster*," *Proceedings of the National Academy of Sciences* 107, no. 46 (2010): 20051–56.

20. J. Morimoto, interview with the author via Zoom, October 8, 2021.

21. C. P. B. Breviglieri and G. Q. Romero, "Acoustic Stimuli from Predators Trigger Behavioural Responses in Aggregate Caterpillars," *Austral Ecology* 44, no. 5 (2019): 880–90.

22. J. H. Hunt et al., "Similarity of Amino Acids in Nectar and Larval Saliva: The Nutritional Basis for Trophallaxis in Social Wasps," *Evolution* 26, no. 6 (1982): 1318–22.

23. Y.-K. Tea et al., "Kleptopharmacophagy: Milkweed Butterflies Scratch and Imbibe from Apocynaceae-Feeding Caterpillars," *Ecology* 102, no. 12 (2021): e03532.

24. Z.-L. Cowan et al., "Predation on Crown-of-Thorns Starfish Larvae by Damselfishes," *Coral Reefs* 35, no. 4 (2016): 1253–62.

25. S. Maslakova, interview with the author via Zoom, April 1, 2021.

26. G. von Dassow et al., "Hoplonemertean Larvae Are Planktonic Predators That Capture and Devour Active Animal Prey," *Invertebrate Biology* 141, no. 1 (2022): e12363.

27. K. Takatsu and O. Kishida, "An Offensive Predator Phenotype Selects for an Amplified Defensive Phenotype in Its Prey," *Evolutionary Ecology* 27, no. 1 (2013): 1–11.

28. M. Weiss, interview with the author via Zoom, November 3, 2021.

29. Ibid.

30. M. R. Weiss, "Good Housekeeping: Why Do Shelter-Dwelling Caterpillars Fling Their Frass?" *Ecology Letters* 6 no. 4 (2003): 361–70.

31. M. Abarca et al., "Host Plant and Thermal Stress Induce Supernumerary Instars in Caterpillars," *Environmental Entomology* 48, no. 1 (2020): 123–31.

CHAPTER 7 | Lessons from Larvae: How Evolution Shaped Development and Vice Versa

1. W. Garstang et al., *Larval Forms, and Other Zoological Verses* (Chicago: University of Chicago Press, 1985).

2. H. de Vries, *Species and Varieties: Their Origin by Mutation* (Chicago: Open Court Publishing, 1904).

3. A. O. Kovalevskij, *Entwicklungsgeschichte des Amphioxus lanceolatus* (St. Petersburg Academie Imperiale des sciences, 1867).

4. E. Haeckel, *Generelle Morphologie Der Organismen. Allgemeine Grundzüge Der Organischen Formen-Wissenschaft, Mechanisch Begründet Durch Die von Charles Darwin Reformirte Descendenztheorie* (Berline: G. Reimer, 1866).

5. W. Garstang, "The Theory of Recapitulation: A Critical Re-Statement of the Biogenetic Law," *Journal of the Linnean Society of London: Zoology* 35, no. 232 (1922): 81–101.

6. A. Ibrahim and M. Gad, "The Occurrence of Paedogenesis in *Eristalis* Larvae (Diptera: Syrphidae)," *Journal of Medical Entomology* 12, no. 2 (June 1975): 268.

7. C. Jaspers et al., "Ctenophore Population Recruits Entirely through Larval Reproduction in the Central Baltic Sea," *Biology Letters* 8, no. 5 (2012): 809–12.

8. T. H. Morgan, "The Rise of Genetics. II," *Science* 76, no. 1970 (1932): 285–88.

9. Garstang, *Larval Forms.*

10. W. McGinnis et al., "Homologous Protein-Coding Sequence in Drosophila Homeotic Genes and Its Conservation in Other Metazoans," *Cell* 37, no. 2 (1984): 403–8.

11. W. McGinnis et al., "Conserved DNA Sequence in Homoeotic Genes of the Drosophila Antennapedia and Bithorax Complexes," *Nature* 308, no. 5958 (1984): 428–33.

12. M. M. Müller et al., "A Homeo-Box-Containing Gene Expressed during Oogenesis in Xenopus," *Cell* 9, no. 1 (1984): 157–62.

13. M. Levine et al., "Human DNA Sequences Homologous to a Protein Coding Region Conserved between Homeotic Genes of Drosophila," *Cell* 38, no. 3 (1984): 667–73.

14. A. E. Carrasco et al., "Cloning of an *X. laevis* Gene Expressed during Early Embryogenesis Coding for a Peptide Region Homologous to Drosophila Homeotic Genes." *Cell* 37, no. 2 (1984): 409–14.

15. A. P. Moczek, interview with the author via Zoom, September 24, 2021.

16. E. M. Standen et al., "Developmental Plasticity and the Origin of Tetrapods," *Nature* 513, no. 7515 (2014): 54–58.

17. W. P. Macdonald et al., "Butterfly Wings Shaped by a Molecular Cookie Cutter: Evolutionary Radiation of Lepidopteran Wing Shapes Associated with a Derived Cut/Wingless Wing Margin Boundary System," *Evolution & Development* 12, no. 3 (2010): 296–304.

18. S. F. Gilbert and D. Epel, *Ecological Developmental Biology: The Environmental Regulation of Development, Health, and Evolution*, Second Edition (New York: Oxford University Press, 2015).

19. A. P. Moczek et al., "When Ontogeny Reveals What Phylogeny Hides: Gain and Loss of Horns During Development and Evolution of Horned Beetles," *Evolution* 60, no. 11 (2006): 2329–41.

20. T.-Y. S. Park and J.-H. Kihm, "Post-Embryonic Development of the Early Ordovician (ca. 480 Ma) Trilobite Apatokephalus Latilimbatus Peng, 1990 and the Evolution of Metamorphosis," *Evolution & Development* 17, no. 5 (2015): 289–301.

21. R. R. Strathmann, "Multiple Origins of Feeding Head Larvae by the Early Cambrian," *Canadian Journal of Zoology* 98, no. 12 (2020): 761–76.

22. R. Strathmann, interview with the author via Zoom, March 5, 2021.

23. T. Miyashita et al., "Non-Ammocoete Larvae of Palaeozoic Stem Lampreys," *Nature* 591, no. 7850 (2021): 408–12.

CHAPTER 8 | Raising Them Right: Conservation and Sustainability

1. J. Burnett, interview with the author via Zoom, October 21, 2021.

2. "California Condor," California Department of Fish and Wildlife, accessed June 1, 2022, wildlife.ca.gov/Conservation/Birds/California-Condor.

3. Burnett, 2021.

4. W. Fialkowski et al., "Mayfly Larvae (*Baetis rhodani* and *B. vernus*) as Biomonitors of Trace Metal Pollution in Streams of a Catchment Draining a Zinc and Lead Mining Area of Upper Silesia, Poland," *Environmental Pollution* 121, no. 2 (2003): 253–67.

5. P. M. Stepanian et al., "Declines in an Abundant Aquatic Insect, the Burrowing Mayfly, across Major North American Waterways." *PNAS* 117, no. 6 (2020): 2987–92.

6. C. W. Twining et al., "Aquatic and Terrestrial Resources Are Not Nutritionally Reciprocal for Consumers," *Functional Ecology* 33, no. 10 (2019): 2042–52.

7. J. K. Tomberlin, interview with the author via Zoom, October 7, 2021.

8. J. K. Tomberlin and A. van Huis, "Black Soldier Fly from Pest to 'Crown Jewel' of the Insects as Feed Industry: An Historical Perspective," *Journal of Insects as Food and Feed* 6, no. 1 (2020): 1–4.

9. W.-M. Wu, interview with the author via Zoom, September 29, 2021.

10. A. M. Brandon et al., "Biodegradation of Polyethylene and Plastic Mixtures in Mealworms (Larvae of *Tenebrio molitor*) and Effects on the Gut Microbiome," *Environmental Science & Technology* 52, no. 11 (2018): 6526–33.

11. Wu, 2021.

12. J. Morimoto, "Addressing Global Challenges with Unconventional Insect Ecosystem Services: Why Should Humanity Care about Insect Larvae?," *People and Nature* 2, no. 3 (2020): 582–95.

13. T. B. Mccormick et al., "Effect of Temperature, Diet, Light, and Cultivation Density on Growth and Survival of Larval and Juvenile White Abalone *Haliotis sorenseni* (Bartsch, 1940)," *Journal of Shellfish Research* 35, no. 4 (2016): 981–92.

14. R. Collin, interview with the author via Zoom, March 29, 2021.

15. F. Á. Fernández-Álvarez et al., "Predatory Flying Squids Are Detritivores during Their Early Planktonic Life," *Scientific Reports* 8, no. 1 (2018): 3440.

CHAPTER 9 | Metamorphosis: But Happier Than Kafka

1. R. Dove, *Mother Love: Poems* (New York: Norton, 1996).

2. S. Gilbert, interview with the author via Zoom, November 18, 2021.

3. S. Maslakova, interview with the author via Zoom, April 1, 2021.

4. N. Morehouse, interview with the author via Zoom, March 25, 2021.

5. Maslakova, 2021.

6. L. N. Vandenberg et al., "Normalized Shape and Location of Perturbed Craniofacial Structures in the Xenopus Tadpole Reveal an Innate Ability to Achieve Correct Morphology," *Developmental Dynamics* 241, no. 5 (2012): 863–78.

7. M. Weiss, interview with the author via Zoom, November 3, 2021.

8. Ibid.

9. R. B. Emlet and O. Hoegh-Guldberg, "Effects of Egg Size on Postlarval Performance: Experimental Evidence from a Sea Urchin," *Evolution* 51, no. 1 (1997): 141–52.

10. N. I. Morehouse et al., "Seasonal Selection and Resource Dynamics in a Seasonally Polyphenic Butterfly," *Journal of Evolutionary Biology* 26, no. 1 (2012): 175–85.

11. Weiss, 2021.

12. Ibid.

13. K. Sláma and C. M. Williams, "The Juvenile Hormone. V. The Sensitivity of the Bug, *Pyrrhocoris apterus*, to a Hormonally Active Factor in American Paper-Pulp," *The Biological Bulletin* 130, no. 2 (1966): 235–46.

14. S. H. Lee et al., "Identification of Plant Compounds That Disrupt the Insect Juvenile Hormone Receptor Complex," *PNAS* 112, no. 6 (2015): 1733–38.

15. M. F. Strathmann and R. R. Strathmann, "An Extraordinarily Long Larval Duration of 4.5 Years from Hatching to Metamorphosis for Teleplanic Veligers of *Fusitriton oregonensis*," *Biological Bulletin* 213, no. 2 (2007): 152–59.

16. C. Lowe, interview with the author via Zoom, April 5, 2021.

17. B. Gaylord et al., "Turbulent Shear Spurs Settlement in Larval Sea Urchins," *PNAS* 110, no. 17 (2012): 6901–6.

18. M. J. A. Vermeij et al., "Coral Larvae Move toward Reef Sounds," *PLOS ONE* 5, no. 5 (2010): e10660.

19. T. A. C. Gordon et al., "Acoustic enrichment can enhance fish community development on degraded coral reef habitat," *Nature Communications* 10, no. 5414 (2019).

20. M. G. Hadfield, interview with the author via Zoom, August 16, 2021.
21. Ibid.
22. Ibid.
23. N. J. Shikuma et al., "Marine Tubeworm Metamorphosis Induced by Arrays of Bacterial Phage Tail–Like Structures," *Science* 343, no. 6170 (2014): 529–33.

CHAPTER 10 | Juveniles: Neither One Thing nor Another

1. L. Hughes, "Youth," poets.org/poem/youth-0.
2. S. Worcester and S. Gaines, "Quantifying Hermit Crab Recruitment Rates and Megalopal Shell Selection on Wave-Swept Shores," *Marine Ecology Progress Series* 157 (1997): 307–10.
3. K. D. Fausch and T. G. Northcote, "Large Woody Debris and Salmonid Habitat in a Small Coastal British Columbia Stream," *Canadian Journal of Fisheries and Aquatic Sciences* 49, no. 4 (1992): 682–93.
4. D. Zacherl, interview with the author (Fullerton, California), July 20, 2021.
5. R. Emlet, interview with the author via Zoom, April 2, 2021.
6. Ibid.
7. A. Moran, interview with the author via Zoom, April 23, 2021.
8. A. Hamdoun, interview with the author via Zoom, July 22, 2021.
9. J. Hodin, interview with the author (Friday Harbor, Washington), December 29, 2021.
10. Ibid.
11. M. Byrne, interview with the author via Zoom, April 19, 2021.
12. Ibid.
13. Morehouse, N, interview with the author via Zoom, March 25, 2021.
14. J. T. Goté et al., "Growing Tiny Eyes: How Juvenile Jumping Spiders Retain High Visual Performance in the Face of Size Limitations and Developmental Constraints," *Vision Research* 160 (2019): 24–36.
15. G. Lingham et al., "How does spending time outdoors protect against myopia? A review," *British Journal of Ophthalmology* 104, no. 5 (2020): 593–99.
16. J. Ward et al., "Why Do Vultures Have Bald Heads? The Role of Postural Adjustment and Bare Skin Areas in Thermoregulation," *Journal of Thermal Biology* 33, no. 3 (2008): 168–73.
17. J. A. Amat and M. A. Rendón, "Flamingo Coloration and Its Significance," *Flamingos: Behavior, Biology, and Relationship with Humans*, M. J. Anderson, ed. (Nova Publishers, 2017): 77–95.
18. J. Burnett, interview with the author via Zoom, October 21, 2021.

CHAPTER 11 | Emergence: A Cicada Case Study

1. G. Kritsky, interview with the author via Zoom, May 13, 2021.
2. Ibid.
3. Ibid.
4. Ibid.
5. G. Williams, interview with the author via Zoom, May 10, 2021.
6. Ibid.

7. D. Gruner, interview with the author (Silver Spring, Maryland), May 21, 2021.

8. Williams, 2021.

9. Ibid.

10. F. Keesing et al., "Hosts as Ecological Traps for the Vector of Lyme Disease," *Proceedings of the Royal Society B: Biological Sciences* 276, no. 1675 (2009): 3911–19.

11. C. Hennessy and K. Hild, "Are Virginia Opossums Really Ecological Traps for Ticks? Groundtruthing Laboratory Observations," *Ticks and Tick-Borne Diseases* 112, no. 5 (2021): 101780.

12. G. Kritsky, *Periodical Cicadas: The Brood X Edition: Black and White Edition* (Ohio Biological Survey, 2021).

EPILOGUE | Our Quiet Dependence on Babies

1. K. Gibran, "On Children," Academy of American Poets, accessed June 1, 2022, poets.org/poem/children-1.

2. T. H. Morgan, "The Rise of Genetics. II," *Science* 76, no. 1970 (1932): 285–88.

3. S. F. Gilbert and D. Epel, *Ecological Developmental Biology: The Environmental Regulation of Development, Health, and Evolution*, Second Edition (New York: Oxford University Press, 2015).

4. J. Tomberlin, interview with the author via Zoom, October 7, 2021.

5. A. Fraser et al., "The Evolutionary Ecology of Age at Natural Menopause: Implications for Public Health," *Evolutionary Human Sciences* 2 (2020): e57.

6. P.-A. Jansson et al., "Probiotic Treatment Using a Mix of Three Lactobacillus Strains for Lumbar Spine Bone Loss in Postmenopausal Women: A Randomised, Double-Blind, Placebo-Controlled, Multicentre Trial," *The Lancet Rheumatology* 1, no. 3 (2019): e154–e62.

7. G. A. Johnson, "A Tumor, the Embryo's Evil Twin," *The New York Times*, March 17, 2014.

8. K. Foltz, interview with the author via Zoom, March 29, 2021.

9. S. F. Gilbert and D. Epel, *Ecological Developmental Biology: The Environmental Regulation of Development, Health, and Evolution*, Second Edition (New York: Oxford University Press, 2015).

10. A. P. Moczek, interview with the author via Zoom, September 24, 2021.

11. J. Tomberlin, interview with the author via Zoom, October 7, 2021.

ACKNOWLEDGMENTS

I wouldn't be here to write this book without the enormous contributions of my parents, Sue and Ben. From mitochondria and milk to lessons and love, their generous generational gifts have been fundamental to my ongoing development. I'm equally grateful to my own children, Ursula and Ulric, for teaching me about the early life stages of humans and for their insight, curiosity, and empathy toward all the other young animals in the world.

Anton, my partner in parenting and the rest of life, has a long arm that's great for photographing baby birds in nests. He's perceptive and patient, even when I won't shut up about parasitic wasps, and he makes the best tea.

Many other friends helped my writing process, notably Nina, who provided feedback and encouragement on an early draft, and Laura, who humored and hosted my cicada obsession. Dozens of scientists, some of whom I've known for years and some of whom I still haven't met in person, shared their time and knowledge. Richard Strathmann, Fernanda Oyarzun, and all the Comparative Invertebrate Embryology crew deserve immense credit for the inspiration behind this book. Scott Gilbert's *Ecological Developmental Biology*, coauthored with David Epel, served as a touchstone throughout the project, and I appreciate both authors taking the time to speak with me. Chris Lowe, Amro Hamdoun, Nick Shikuma, and Jason Hodin welcomed me into their labs, and Danielle Zacherl, into her home. Many others were equally charitable over

Zoom, even when time zones forced us to meet at odd hours. I'm sorry I can't list every name, but I am deeply grateful for every conversation and correction. Mistakes that remain are mine alone.

My splendid agent, Stacey Kondla, and marvelous editors, Matthew Lore and Nicholas Cizek, helped nurture and shape this book through its own diverse developmental stages. Many thanks to the whole team at The Experiment and to Javier Lazaro for the exquisite jacket art. For visuals inside the book, I'm indebted to Rob Lang as well as to the talent and kindness of numerous other artists and photographers.

I was fortunate to have the opportunity to write and research at the Helen Riaboff Whiteley Center of Friday Harbor, Washington, on the same rich ground where my love of larvae first sprouted.

This is a book about birth and rebirth, joy, and hope, and creation. Thus, cheesy as it may be, I extend my thanks to every beautiful beginning in our world—from the gluttonous gastrula to the chattering chick to the newest infant human.

IMAGE CREDITS

Photo Insert
1: Frank Baensch
3 and 8: Nature Picture Library/Alamy Stock Photo
4: Greg Rouse
6: Reimar/Adobe Stock
7: Darlyne Murawski
9: Thomas van de Kamp
10: Linda Ianniello
11: Juliano Morimoto
12: Australian Museum/Wikimedia Commons
13: Alan Henderson/Cover Images
14: Marlin Harms
15: Jerry Kirkhart
17: Jeremy Brown/Dreamstime.com

INDEX

NOTE: Page numbers in *italics* refer to figures and photos; page numbers in italics and followed by *in* refer to photo insert.

ABOUT THE AUTHOR

DANNA STAAF earned a PhD in invertebrate biology from Stanford University with her studies of baby squid. She is the author of *Monarchs of the Sea: The Extraordinary 500-Million-Year History of Cephalopods* and *The Lady and the Octopus: How Jeanne Villepreux-Power Invented Aquariums and Revolutionized Marine Biology*, and she has written for *Science*, *Smithsonian*, *Atlas Obscura*, and *Nautilus*. She lives in California with her human family, a cat, and a garden full of grubs, caterpillars, maggots, and innumerable other babies.

dannastaaf.com | @dannastaaf | ⌾ dannajoystaaf